Theoretical Physics for Biological Systems

Theoretical Physics for Biological Systems

Paola Lecca

Researcher

Department of Mathematics,
University of Trento, Italy

Angela Re

Postdoctoral Fellow

Center for Sustainable Future Technologies CSFT@Polito,
Istituto Italiano di Tecnologia, Italy

CRC Press
Taylor & Francis Group
Boca Raton London New York

CRC Press is an imprint of the
Taylor & Francis Group, an **informa** business

A SCIENCE PUBLISHERS BOOK

Cover illustration: In the cover image, the multiple copies of Schrödinger's cat represent the multiplicity of states of a physical system. The snout of the cat is marked with the Ψ letter, which conventionally denotes the wave function. Schrödinger's cats surround a strand of DNA to represent the presence of quantum effects in biological processes.

The authors thank Michela Lecca, Researcher at Fondazione Bruno Kessler of Trento, Italy, and painter for hobby, who imagined and created this representation.

CRC Press
Taylor & Francis Group
6000 Broken Sound Parkway NW, Suite 300
Boca Raton, FL 33487-2742

First issued in paperback 2020

© 2019 by Taylor & Francis Group, LLC
CRC Press is an imprint of Taylor & Francis Group, an Informa business

No claim to original U.S. Government works

ISBN-13: 978-1-138-55241-8 (hbk)
ISBN-13: 978-0-367-78038-8 (pbk)

Visit the Taylor & Francis Web site at
http://www.taylorandfrancis.com

and the CRC Press Web site at
http://www.crcpress.com

Preface

The long-lasting commonplace that physics is the science that studies inert matter while biology is the science that studies living matter have fallen into disuse for several years. Even the misconception that physics adopts a reductionist approach to the study of complex systems has now lost its meaning.

The physics that was the first discipline to make a massive use of the language of mathematics defined a methodological protocol for the study of complex systems that was recently also adopted by systems biology and which led to the birth of computational biology. The use of ordinary differential equations for the modelling of network dynamics, the use of advanced statistical concepts for the inference of models and parameters of biological process models as well as for the interpretation of experimental data is now widespread in biology.

However, in order to build mathematical models of natural phenomena, physics and in particular theoretical physics are not restricted to the aforementioned tools but employ tools and concepts of mathematical logic, algebra, general topology, analysis and measurement theory and differential geometry. Theoretical physics uses these tools to describe mathematically the representation of an observable phenomenon or of a conjecture. The contribution brought by theoretical physics to systems biology is twofold: on the one hand it offers abstract representations that are at the base of models or that can be useful for validating a conjecture, and, on the other it identifies the mathematical tools with which to describe these representations. Modern biology is essentially systems biology, but, above all, it is a quantitative discipline. This new role of biology allows the convergence of mathematics, physics and life sciences in general.

This book focuses on two branches of theoretical physics that prove promising in the description of the static and dynamic properties of biological systems: quantum physics and statistical mechanics. Recent experiments have confirmed the occurrence of quantum phenomena in many biological processes, but, above all, quantum physics and statistical mechanics inspire computational models of network dynamics at each scale.

The book is addressed both to those readers with a mathematical and physical background facing the world of modern systems biology, and to those readers with a biological and/or computational background who wish to explore and take inspiration from the models of theoretical physics. The book does not want to be a textbook of quantum biology, nor does it want to deepen specific case studies, but rather to provide the reader with the basic concepts of quantum physics and statistical mechanics of useful application in the computational systems biology. The first Chapter introduces the reader to quantum mechanics and presents the most common biological processes and systems governed by quantum phenomena. The second Chapter introduces the concepts of statistical mechanics that have become the theoretical foundations of the modern biological network simulation algorithms. The third Chapter presents the measures of network centrality that are inspired by models of physical systems. The fourth Chapter presents the application contexts of these centrality measures and the best known algorithms of quantum computating, currently implemented by many IT systems, and which have recently proved to operate in many processes regulating the energetic efficiency of various biological systems. The fifth Chapter summarizes the motivations that are leading towards an increasingly fruitful convergence between theoretical physics and systems biology and figures out some future perspectives.

The book proposes thirty exercises for the self-assessment of the understanding of the its contents, and useful also for didactic purposes. The level of mathematical knowledge required to understand the contents of the book and solve the exercises is that provided in the basic university courses common to the various scientific disciplines.

Paola Lecca
Angela Re

Contents

List of Figures

List of Tables

Chapter 1

Quantum Mechanics in Biology

The quantum mechanics is the science of the very small at the scale of the energy levels of atoms and subatomic particles. The majority of the biological processes involving the conversion of energy into forms that are usable for chemical transformations are quantum mechanical in nature. In this chapter we present the basic concepts and postulates of quantum mechanics, and some well-known biological processes governed by quantum physics.

1.1 General Definitions

In quantum mechanics the state of a system (e.g. an ensemble of particles such as atoms or molecules) is described by a wave function Ψ. For a system of N particles, the wave function depends on the coordinates $x_1, x_2, \cdots, x_N$ of all the particles and on the time t. Ψ is defined on a complex Hilbert space and its squared module $|\Psi|^2$ multiplied by the volume element $d\mathbf{v} = dx_1 d_2, \ldots x_N$ expresses the probability of finding a particle i between x_i and $x_i + dx_i$ ($i = 1, 2, \ldots, N$). If Ψ is normalized (i.e. it represents a density of probability), then

$$\int \Psi^* \Psi d\mathbf{v} = 1 \tag{1.1}$$

where Ψ^* is the complex conjugate of Ψ. The Eq.1.1 states that the likelihood of finding the particle in the volume of space in which it can actually move is maximum.

The time evolution of the system of particles is governed by the Schrödinger equation

$$i\hbar \frac{\partial \Psi}{\partial t} = H\Psi \qquad (1.2)$$

where $h = 6.63 \cdot 10^{-34}$ J·s is the Planck constant, $\hbar = \frac{h}{2\pi}$, and H is the Hamiltonian operator. H is associated with the total energy of the system, i.e. the sum of kinetic energy (that of motion) and the potential energy (that of position). In one dimension, the Hamiltonian operator for a particle of mass m is

$$H = -\frac{\hbar^2}{2m} \frac{\partial^2}{\partial x^2} + V(x)$$

where $-\frac{\hbar}{2m} \frac{\partial}{\partial x}$ is the quantum analogue to the classical variable *momentum* of the particle.

The Hamiltonian is a linear Hermitian operator, i.e. give two wave functions Ψ_1 and Ψ_2

$$H(c_1\Psi_1 + c_2\Psi_2) = c_1 H\Psi_1 + c_2 H\Psi_2$$

for any constants c_1 and $c_2 \in \mathbb{C}$, and

$$\int \Psi^* H\Phi d\mathbf{v} = \int \Phi (H\Psi)^* d\mathbf{v} \qquad (1.3)$$

where Ψ and Φ are any twice differentiable functions for which $\int \Psi^*\Psi d\mathbf{v}$ and $\int \Phi^*\Phi d\mathbf{v}$ are finite. Traditionally, the integrals in Eq. (1.3) are written in the Dirac brackets notations as in the following.

$$\langle \Psi | H\Phi \rangle = \langle H\Psi | \Phi \rangle$$

Finally, for a system governed by a wave function Ψ, the average or the expectation value of any dynamical variable represented by a quantum operator G is

$$\langle G \rangle = \frac{\langle \Psi | G | \Psi \rangle}{\langle \Psi | \Psi \rangle} \qquad (1.4)$$

where $\langle \Psi | \Psi \rangle = \int \Psi^*\Psi d\mathbf{v}$.

1.2 The Time-independent Schrödinger Equation

If the Hamiltonian is time independent, the Schrödinger equation can be separated into coordinate- and time- dependent parts. Writing

$$\Psi(x_1, x_2, \ldots, x_N, t) = \varphi(x_1, x_2, \ldots, x_N)\chi(t) \tag{1.5}$$

we can separate the variables in Eq. (1.2) to obtain

$$-i\hbar\frac{d\varphi}{dt} = E\varphi \tag{1.6}$$

whose solution is

$$\chi(t) = e^{-i\frac{Et}{\hbar}} \tag{1.7}$$

and the so called time-independent Schrödinger equation

$$H\varphi = E\varphi. \tag{1.8}$$

E is a separation constant whose physical significance is that of an energy. In fact, the time-independent Schrödinger equation, the operation of the Hamiltonian on the wave function produces specific values of the energy, called *eigenvalues* of H. The wave function to which the Hamiltonian operator is applied is called *eigenfunction* of H.

It is straightforward to prove that the eigenvalue E is the expectation value of the Hamiltonian of the system, i.e. $E = \langle H \rangle$. Using Eq. (1.5), (1.7), and (1.8), we can evaluate Eq. (1.4) for H:

$$\langle H \rangle = \frac{\langle \Psi | H | \Psi \rangle}{\langle \Psi | \Psi \rangle} = \int (\phi^* \chi^*) H(\phi\chi)d\mathbf{v}$$
$$= \int (\phi^* \chi^*) E(\phi\chi)d\mathbf{v} = E \int \phi^* \phi \, d\mathbf{v} = E.$$

In many real problems in biochemistry, the solution of time independent Schrödinger equation is hard to obtain exactly. A good approximation can be developed the potential V is time independent and very small. In this case, we can write the Hamiltonian as the sum of a zeroth-order Hamiltonian H_0 and a perturbation λV, where λ is a parameter quantifying the order of the perturbation:

$$H = H_0 + \lambda V \tag{1.9}$$

Let $\phi_n^{(0)}$ the eigenfunctions of H_0 and E_n^0 its eigenvalues. The eigenfunction ψ of H can then be written as

$$\psi_n = \phi_n^{(0)} + \lambda\phi_n^{(1)} + \lambda^2\phi_n^{(2)} + \ldots \tag{1.10}$$

and the eigenvalues of H can be written as

$$E_n = E_n^{(0)} + \lambda E_n^{(1)} + \lambda^2 E_n^{(2)} + \dots. \tag{1.11}$$

Therefore

$$H\psi_n = (H_0 + \lambda V)(\phi_n^{(0)} + \lambda \phi_n^{(1)} + \lambda^2 \phi_n^{(2)} + \dots)$$
$$= H_0 \phi_n^{(0)} + \lambda(H_0 \phi_n^{(1)} + V\phi_n^{(0)}) + \lambda^2(H_0 \phi_n^{(2)} + V + \phi_n^{(1)}) + \dots$$

and, equating like powers of λ to the terms of Eq. (1.11) we obtain:

$$H_0 \phi_n^{(0)} = E_n^{(0)} \phi_n^{(0)} \tag{1.12}$$

$$H_0 \phi_n^{(1)} + V\phi_n^{(0)} = E_n^{(0)} \phi_n^{(1)} + E_n^{(1)} \phi_n^{(0)} \tag{1.13}$$

$$H_0 \phi_n^{(2)} + V\phi_n^{(1)} = E_n^{(0)} \phi_n^{(2)} + E_n^{(2)} \phi_n^{(1)} \tag{1.14}$$

$$\dots$$

The j-th term is

$$H_0 \phi_n^{(j)} + V\phi_n^{(j-1)} = \sum_{k=1}^{j} E_n^{(j)} \phi_n^{(j-1)}.$$

The equation (1.12) is satisfied by assumption. The solution of the Eq. (1.13) can be obtained by multiplying the left- and right-hand side of the equation by $(\phi_k^{(0)})^*$, expressing $\phi_n^{(1)}$ in terms of the zeroth-order state, i.e.

$$\phi_n^{(1)} = \sum_{k \neq n} c_{nk}^{(1)} \phi_k^{(0)}, \tag{1.15}$$

and integrating, as in the following.

$$V\phi_n^{(0)} + H_0 \phi_n^{(1)} = E_n^{(1)} \phi_n^{(0)} + E_n^{(0)} \phi_n^{(1)}$$
$$(\phi_k^{(0)})^* (V\phi_n^{(0)} + H_0 \phi_n^{(1)}) = (\phi_k^{(0)})^* (E_n^{(1)} \phi_n^{(0)} + E_n^{(0)} \phi_n^{(1)})$$
$$(\phi_k^{(0)})^* V\phi_n^{(0)} + (\phi_k^{(0)})^* H_0 \phi_n^{(1)} = (\phi_k^{(0)})^* E_n^{(1)} \phi_n^{(0)} + (\phi_k^{(0)})^* E_n^{(0)} \phi_n^{(1)})$$

$$\int (\phi_k^{(0)})^* E_n^{(1)} \phi_n^{(0)} d\mathbf{v} + \int (\phi_k^{(0)})^* E_n^{(0)} \phi_n^{(1)}) d\mathbf{v} =$$
$$\int (\phi_k^{(0)})^* E_n^{(1)} \phi_n^{(0)} d\mathbf{v} + \int (\phi_k^{(0)})^* E_n^{(0)} \phi_n^{(1)}) d\mathbf{v}.$$

Let us assume that

$$\int (\phi_n^{(0)})^* \phi_n^{(1)} d\mathbf{v} = 0$$

then, if $k = n$ we obtain

$$\int (\phi_n^{(0)})^* \, E_n^{(1)} \phi_n^{(0)} d\mathbf{v} = \int (\phi_n^{(0)})^* \, E_n^{(1)} \phi_n^{(0)} d\mathbf{v} d\mathbf{v}.$$

so that, in the Dirac parenthesis notation,

$$E_n^{(0)} = \langle \phi_n^{(0)} | V | \phi_n^{(0)} \rangle. \tag{1.16}$$

If $k \neq n$, we obtain

$$c_{nk}^{(1)} = \frac{\langle \phi_k^{(0)} | V | \phi_n^{(0)} \rangle}{E_n^{(0)} - E_k^{(0)}}. \tag{1.17}$$

In similar way, we can develop expressions for the higher-order terms in both the energy and the wave functions [193].

An important application of the time-independent Schrödinger equation is the quantum harmonic oscillator. Indeed, in quantum dynamics, as in classical one, the harmonic oscillator is the simplest system in which attractive forces are present. As a consequence, the harmonic oscillator represents an important paradigm for the majority of vibrational phenomena.

The Schrödinger equation for a one dimensional harmonic oscillator is

$$\frac{d^2}{dx^2} \psi = -\frac{2m}{\hbar^2} \left(E_{kinetics} - \frac{1}{2} K x^2 \right) \tag{1.18}$$

$E_{kinetics}$ is the kinetic energy of the particle, and K is the force constant (the force of the mass being $F = -Kx$ proportional to the position x and directed toward the origin).

The angular frequency of a classical oscillator is

$$\omega = \sqrt{\frac{K}{M}} \tag{1.19}$$

As suggested by Schatz et al. [193], in order to simplify the notation it is convenient to adopt adimensional units. To this purpose, we introduce the adimensional variable

$$\xi = \sqrt{\frac{m\omega}{\hbar}} \cdot x. \tag{1.20}$$

using Eq. (1.19) for ω, and the variable ϵ defined as

$$\epsilon \equiv \frac{E}{\hbar \omega}, \tag{1.21}$$

inserting Eq. (1.20) and Eq. (1.21) into Eq. (1.18), we have that

$$\frac{m\omega}{\hbar}\frac{d^2}{d\xi^2}\psi(\xi) = -\frac{2m}{\hbar^2}\left(E - \frac{1}{2}K\frac{\hbar}{m\omega}\xi^2\right)\psi(\xi)$$

$$\frac{d^2}{d\xi^2}\psi(\xi) = -\frac{2}{\hbar\omega}\left(E - \frac{1}{2}K\frac{\hbar}{m\omega}\xi^2\right)\psi(\xi),$$

and finally

$$\frac{d^2}{d\xi^2}\psi(\xi) = -2\left(\epsilon - \frac{1}{2}\xi^2\right)\psi(\xi). \tag{1.22}$$

For large ξ, ϵ can be neglected and the solution of the Eq. (1.22) is

$$\psi(\xi) \approx \xi^n e^{-\frac{\xi^2}{2}}$$

where n takes any finite value. More generally, we can write the solution of the Eq. (1.22) as

$$\psi(\xi) = f(\xi)e^{-\frac{\xi^2}{2}} \tag{1.23}$$

where $f(\xi)$ is a polynomial function that must not grow like e^{ξ^2} (in that case, ψ would not have any physical meaning[1]).

Substituting Eq. (1.23) in Eq. (1.22), we have that

$$\frac{d^2}{d\xi^2}\left(f(\xi)e^{-\frac{\xi^2}{2}}\right) = -2\left(\epsilon - \frac{\xi^2}{2}\right)f(\xi)e^{-\frac{\xi^2}{2}}$$

$$\frac{d}{d\xi}\left(f'(\xi)e^{-\frac{\xi^2}{2}} - f(\xi)\xi e^{-\frac{\xi^2}{2}}\right) = f''(\xi)e^{-\frac{\xi^2}{2}} - 2f'(\xi)\xi e^{-\frac{\xi^2}{2}} - f(\xi)e^{-\frac{\xi^2}{2}}$$

where $f'(\xi) = \frac{d\psi(\xi)}{d\xi}$ and $f''(\xi) = \frac{d^2\psi(\xi)}{d\xi^2}$. Therefore,

$$f''(\xi) - 2\xi f'(\xi) - (1 - \xi^2)f(\xi) = -2\left(\epsilon - \frac{\xi^2}{2}\right)f(\xi)$$

$$f''(\xi) - 2\xi f'(\xi) = (1 - 2\epsilon)f(\xi),$$

and finally, we obtain

$$f''(\xi) - 2\xi f'(\xi) - (2\epsilon - 1)f(\xi) = 0 \tag{1.24}$$

---

[1] We let the reader demonstrate that $f(\xi)$ must not grow like e^{ξ^2}. This can be easily proven by showing that $\psi(\xi) = e^{\frac{\xi^2}{2}}$ is a solution of Eq. (1.22) without physical meaning.

If we expand the polynomial function $f(\xi)$ into a series (in principle infinite)

$$f(\xi) = \sum_{n=0}^{\infty} A_n \xi^n \tag{1.25}$$

so that its first and second derivatives are:

$$f'(\xi) = \sum_{n=0}^{\infty} A_n n \xi^{n-1}$$

$$f''(\xi) = \sum_{n=0}^{\infty} A_{n+1}(n+1)(n+2)\xi^n$$

we have that

$$\sum_{n=0}^{\infty} A_{n+1}(n+1)(n+2)\xi^n - 2\sum_{n=0}^{\infty} A_n n \xi^n + (2\epsilon - 1)\sum_{n=0}^{\infty} A_n \xi^n,$$

and finally

$$\sum_{n=0}^{\infty} \left[A_{n+2}(n+1)(n+2) + A_n(2\epsilon - 2n - 1) \right]\xi^n = 0. \tag{1.26}$$

This equation can be satisfied for any value of ξ only if

$$A_{n+2}(n+1)(n+2) + A_n(2\epsilon - 2n - 1) = 0 \tag{1.27}$$

The Eq. (1.27) defines a recursion relation between the coefficients $A_{(\cdot)}$. If A_0 and A_1 are given, the Eq. (1.27) allows to find by recursion the solution of the Eq. (1.26) in the form of a power series. From Eq. (1.27) we see that for large n, the coefficients of the series behave like

$$\frac{A_{n+2}}{A_n} \longrightarrow \frac{2}{n}$$

that is

$$A_{n+2} \sim \frac{1}{\left(\frac{n}{2}\right)!}. \tag{1.28}$$

The coefficients of

$$e^{\xi^2} = \sum_{n=0}^{\infty} \frac{\xi^{2n}}{n!}$$

behave like Eq. (1.28), so that the recursion relation Eq. (1.27) between the coefficients gives a function $f(\xi)$ growing like e^{ξ^2}, that bring to non-physical diverging solutions. The only condition that allows us to have physically meaningful solutions is that all the coefficients beyond a given n vanish. In this case the infinite series reduces to a finite-degree polynomial. This happens if and only if

$$\epsilon = n + \frac{1}{2} \tag{1.29}$$

where $n \in \mathbb{N}$. Therefore, from Eq. (1.21), we see that allowed energies for the quantum harmonic oscillator are *quantized*:

$$E_n = \left(n + \frac{1}{2} \right) \hbar \omega. \tag{1.30}$$

The corresponding polynomials $f_n(\xi)$ are known as *Hermite polynomials*. $f_n(\xi)$ is of degree n in ξ, is even for even n, is odd for odd n. Since $e^{-\frac{\xi^2}{2}}$ is node-less and even, the complete wave function corresponding to the energy E_n

$$\psi_n(\xi) = f_n(\xi) e^{-\frac{\xi^2}{2}} \tag{1.31}$$

has n nodes and the same parity as n. Finally, we notice that the fact that all solutions of the Schrödinger equation are either odd or even functions is a consequence of the symmetry of the potential ($V(-x) = V(x)$).

1.3   Time-dependent Schrödinger Equation

Up to this point, we have considered the time-independent solution of the Schrödinger equation. However, biology and chemistry are concerned with how the solution evolves in time [37]. Relevant examples are (i) the interaction between matter and radiation, i.e. how atoms and molecules respond to electromagnetic radiation in phenomena like light absorption, and scattering; (ii) chemical reactions, i.e. reactions between atoms and molecules; (iii) intramolecular energy transfer, i.e. intramolecular electron transfer, radiationless transition. These processes are studied by the time-dependent quantum mechanics. In the majority of the time-dependent problems, the systems are supposed to be in an initial stationary state, then a time-dependent interaction is turned on and the systems starts to evolve from its initial state.

Let H_0 be the time-independent initial Hamiltonian and $V(x,t)$ the interaction potential, that is a function of the spatial coordinates x and the time t. The time evolution of the systems is determined by the solution of the time-dependent Schrödinger equation

$$i\hbar \frac{\partial \psi(x,t)}{\partial t} = H\phi(x,t) = (H_0 + V(x,t))\psi(x,t). \tag{1.32}$$

Let $\phi_n(x)$, $(n = 1, 2, \dots)$ the stationary states prior the activation of the interaction potential $V(x, t)$. $\Phi_n(x)$ satisfy the equation

$$H_0 \phi_n = E_n \phi_n \tag{1.33}$$

and have a time dependent part $e^{-\frac{iE_n t}{\hbar}}$. Then

$$\psi(x, t) = \sum_n c_n(t)\phi_n(x)e^{-\frac{iE_n t}{\hbar}} \tag{1.34}$$

where the c_n's are coefficients that can be determined in the following way. We substitute Eq. (1.34) in (1.32), so that we obtain

$$i\hbar\frac{\partial \psi(x, t)}{\partial t} = \sum_n c_n(t)(H_0 + V(x, t))\phi_n(x)e^{-\frac{iE_n t}{\hbar}}$$
$$= \sum_n c_n(t)(E_0 + V(x, t))\phi_n(x)e^{-\frac{iE_n t}{\hbar}}. \tag{1.35}$$

Since we also have that

$$i\hbar\frac{\partial \psi(x, t)}{\partial t} = i\hbar \sum_n \left[\frac{dc_n}{dt} - \frac{i}{\hbar}E_n c_n\right]\phi_n(x)e^{-\frac{iE_n t}{\hbar}} \tag{1.36}$$

we obtain that

$$i\hbar \sum_n \left[\frac{dc_n}{dt} - \frac{i}{\hbar}E_n c_n\right]\phi_n(x)e^{-\frac{iE_n t}{\hbar}} = \sum_n c_n(t)(E_0 + V(x, t))\phi_n(x)e^{-\frac{iE_n t}{\hbar}}. \tag{1.37}$$

Multiplying by $\phi_n^*(x)$ and integrating, we have that

$$i\hbar\frac{dc_n}{dt}e^{-\frac{iE_n t}{\hbar}}\langle k|n\rangle = \sum_n e^{-\frac{iE_n t}{\hbar}}c_n(t)\langle k|V|n\rangle. \tag{1.38}$$

Since $\langle k|n\rangle = \delta_{kn}$ and introducing $V_{kn}(t) = \langle k|V(x, t)|n\rangle$, the Eq. (1.38) becomes

$$\frac{dc_n}{dt} = \frac{i}{\hbar}\sum_n e^{-\frac{iE_n t}{\hbar}}. \tag{1.39}$$

Eq. (1.39) together with the boundary condition $c_n(t = 0) = \delta_{kn}$ uniquely define the solution of the time-dependent Schrödinger equation (1.32). If $E_n - E_k$ and $V_{kn}(t)$ are known, this equation can be numerically solved. However, this information is often missing, and for this reason, the development of approximated solutions is very useful [193].

1.4   Transition Probability Per Unit of Time

Suppose that $\psi(x,0) = \phi_m(x)$. If V_{kn} is small enough so that the change in $c_n(t)$ is small, $c_n(t)$ in Eq. (1.39) can be considered approximately equal to its initial value δ_{kn} along all the time. Therefore, the Eq. (1.39) becomes:

$$\frac{dc_k^{(1)}}{dt} = -\frac{i}{\hbar} V_{kn}(t) e^{\frac{i(E_k - E_n)t}{\hbar}} \tag{1.40}$$

and its solution (in the first order approximation) is

$$c_k^{(1)}(t) \approx c_k(t = 0) - \frac{i}{\hbar} \int_0^t V_{kn}(\tau) e^{\frac{i(E_k - E_m)\tau}{\hbar}} d\tau$$

$$= \delta_{kn} - \frac{i}{\hbar} \int_0^t V_{kn}(\tau) e^{\frac{i(E_k - E_m)\tau}{\hbar}} d\tau. \tag{1.41}$$

The probability that the systems is in the state k at time t, for $k \neq m$ is

$$P_k = |c_k^{(1)}(t)|^2 = \frac{1}{\hbar^2} \left| \int_0^t V_{kn}(\tau) e^{\frac{i(E_k - E_m)\tau}{\hbar}} d\tau \right|^2. \tag{1.42}$$

Let us consider the simple case of a constant interaction $V(t)$ that is turned on at time $t = 0$ and turned off at time $t = T$. An example of this situation in nature are the radiationless transitions from photoexcited states.

The particles of energy are behaving like waves. If $k \neq m$, from Eq. (1.41), we have

$$c_k = -\frac{i}{\hbar} \int_0^T V_{km} e^{i v_{kn}\tau} d\tau \tag{1.43}$$

where $v_{km} = \frac{E_k - E_n}{\hbar}$. If V_{km} is constant over time, then

$$c_k = -V_{km} \frac{\omega_{km} T - 1}{\hbar v_{km}}. \tag{1.44}$$

The probability of being in state k is thus

$$P_k^m = |c_k|^2 = 4|V_{km}|^2 \frac{\sin^2\left[\frac{v_{km}\tau}{2\hbar}\right]}{v_{km}^2}. \tag{1.45}$$

1.5   Quantum Coherence and Entanglement

Quantum coherence arises from quantum superposition and represents a fundamental signature of the departure of quantum mechanics from classical physics [203, 225, 234]. The *superposition principle* asserts that a system is in all possible states at the same time, until it is measured. After measurement it then falls to one of the basis states that form the superposition, thus destroying the original "mixed" configuration. *Coherence* generally refers to all properties of the correlation between several waves or wave packets. The simplest configuration for the study of correlations is that of quantum system consisting of two subsystems A and B, each associated with a (finite dimensional) Hilbert space $\mathcal{H}_A$ and $\mathcal{H}_B$, respectively. Suppose that the system is prepared in a pure quantum state $|\psi\rangle_{AB}$ living in the composite Hilbert space $\mathcal{H}_{AB}$, defined as the tensor product of $\mathcal{H}_A$ and $\mathcal{H}_B$, i.e.

$$\mathcal{H}_{AB} = \mathcal{H}_A \otimes \mathcal{H}_B.$$

By "pure quantum state", we mean a quantum state that cannot be written as a mixture of other states. Two possibilities can occur [1]. The first possibility is that the two subsystems are completely independent, in which case the state takes the form of a tensor product state

$$|\psi\rangle_{AB} = |\alpha\rangle_A \otimes |\beta\rangle_A$$

with $|\alpha\rangle_A \in \mathcal{H}_A$ and $|\beta\rangle_B \in \mathcal{H}_B$. In this case there are no correlations of any form (classical or quantum) between the two subsystems A and B.

The second possibility is that,

$$|\psi\rangle_{AB} \neq |\alpha\rangle_A \otimes |\beta\rangle_A.$$

In this case, $|\psi\rangle_{AB}$ describes an *entangled* state of the two subsystems A and B. Entanglement refers to any possible form of correlation in pure bipartite states, and can manifest themselves in different yet equivalent ways [1]. For example, every pure entangled state can be *nonlocal*. Nonlocality is the phenomenon by which measurements made at a microscopic level contradict the *local realism*. The concept of local realism combines *realism* and *locality*. Realism refers to the idea that nature exists independently of man's mind. In other terms, even if the result of a measurement does not exist before the act of measuring, that does not mean this result is a creation of the mind of the observer. Realism contradicts the "consciousness causes collapse" theory in quantum mechanics. The principle of locality states that an object is influenced only by its close surroundings. It contradicts the principle of *instantaneous* action at distance. Similarly, every pure entangled state can be disturbed by any local measurement. As a consequence, nonlocality, entanglement and quantum correlations are synonymous for pure bipartite states.

Let us now consider a more complex situation in which A and B are globally prepared in a mixed state described by a density matrix $\rho_{AB} \in \mathcal{D}_{AB}$, where $\mathcal{D}_{AB}$ is the convex set of all the operators acting on $\mathcal{H}_{AB}$. Indeed, the set of quantum states consists of density matrices which are Hermitian, positive and normalized by the trace condition [26]. In the density matrix formalism, quantum correlations appear as off-diagonal elements[2].

The state described by the matrix ρ_{AB} is *separable* (or *unentangled*) if it can be expressed in the following form

$$\rho_{AB} = \sum_i p_i \xi_A^{(i)} \otimes \tau_B^{(i)} \tag{1.46}$$

where p_i is a probability distribution, and $\xi_A^{(i)}$ and $\tau_B^{(i)}$ are quantum states. So, the set of separable states $\mathcal{S}_A \in \mathcal{D}_{AB}$ is defined as follows

$$\mathcal{S}_{AB} := \left\{ \rho_{AB} | \rho_{AB} = \sum_i p_i \xi_A^{(i)} \otimes \tau_B^{(i)} \right\} \tag{1.47}$$

Any other state $\rho_{AB} \notin \mathcal{S}_{AB}$ is *entangled*. Hence, any mixed entangled state is defined as the state that cannot be decomposed as a convex mixture of product states. The set of product states $\mathcal{P}_{AB}$ is defined as

$$\mathcal{P}_{AB} = \left\{ \rho_{AB} | \rho_{AB} \rangle \rho_A \otimes \rho_B \right\} \tag{1.48}$$

and in general is strictly smaller than the set of separable states $\mathcal{S}_{AB}$.

In summary, the definition of entanglement has two ingredients: the superposition principle and the tensorial structure of Hilbert space. Within the set of entangled states some other layers of most stringent form of non-classicality can be distinguished: (i) the steerable entangled states, and (ii) nonlocal steerable entangled states. *Steering* is the possibility of manipulating the state of one subsystem by making measurements on the other. Steering is a form of *asymmetric* correlation [1] to express the fact that some states can be steered from A to B but not the other way around, whereas nonlocality is a form of *symmetric* correlation because it is invariant under the swap of the subsystems A and B. Nonlocality and steering can be considered as the most radical departures from classicality, because nonlocality violate the local realism and steering is the essence of the inseparability.

---

[2]However, the number, location and value of these elements depend on the chosen basis vector set. Favouring one basis vector set over the others often results in the preferred basis problem. As a result, coherence cannot be uniquely defined by using off-diagonal elements.

Quantum correlations are stronger than the classical statistical correlations. Figure 1.1 depicts the hierarchy of correlations of composite quantum systems.

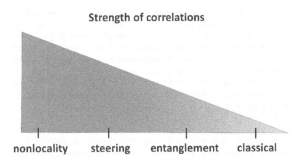

Strength of correlations

nonlocality   steering   entanglement   classical

Figure 1.1: In order of decreasing strength the correlations of composite system are: nonlocality, steering, entanglement, classical statistical correlations.

1.6 Quantum Interference

Quantum interference is one of the most challenging principles of quantum theory. Essentially, the concept states that elementary particles cannot only be in more than one place at any given time through superposition, but that an individual particle, such as a photon can cross its own trajectory and interfere with the direction of its path. Quantum interference is a direct consequence of the coherent superposition of quantum probability fields, just as in classical wave mechanics interference arises from the coherent superposition of energy fields. When more than one particle interacts, the corresponding multi-particle coherent superpositions are described by nonfactorizable (or entangled) states, which cause multi-particle interference processes.

In addition to experiments of molecular interferometry (known to most), quantum interference plays a fundamental rule in experiments based on coherence control, where interference mechanisms are considered in order to enhance or inhibit the formation of products in chemical reactions (see for example the work of Jambrina et al. [114]). Another research field in biology where interference is very important is the selectivity mechanism in ion channels since interference happens between similar ions through the same size of ion channels [188]. For a comprehensive review of the research field and application domains relying of quantum interference, we refer the reader to the paper of Sanz et al. [191]. Here limiting ourselves to the context and the applicative domains presented in this book, it is worth to mention that the theoretical grounds of the quantum information theory (and quantum computation [119]) also rely on the idea of a coherent superposition: the *qubit*. The unit of classical information is the bit, which can have one out of two possible outcomes, 0 or 1, just the most standard binary notation. The quantum-mechanical analogues are the special orthonormal basis states $|0\rangle$ and $|1\rangle$, known as *computational basis*.

According to the quantum mechanics, any state vector represents a realizable physical state. When the state is represented by a linear superposition, we can find the quantum interference between the superposed states $|\psi_1\rangle$ and $|\psi_2\rangle$ as follows:

$$c|\langle x|(|\psi_1\rangle + |\psi_2\rangle)|^2 = c\left[|\langle x|\psi_1\rangle|^2 + |\langle x|\psi_2\rangle|^2 + 2\mathrm{Re}\{\langle x|\psi_1\rangle\langle x|\psi_2\rangle^*\}\right] \quad (1.49)$$

where c is a normalization constant. The third term represents the quantum interference and corresponds to the fact that the quantum probability is affected by off-diagonal elements of the density operator of a coherent superposition state. On the other hand, in the quantum measurement process, if the measurement process itself generates the superposition effect from a standard basis $\{|y\rangle\}$, we have that

$$c|(\langle y| + \langle \delta y|)|\psi\rangle|^2 = c\left[|\langle y|\psi\rangle|^2 + |\langle \delta y|\psi\rangle|^2 + 2\mathrm{Re}\{\langle y|\psi\rangle\langle \delta y|\psi\rangle^*\}\right] \quad (1.50)$$

then the resulting interference term represents the macroscopic quantum interference effect by the quantum measurement itself.

1.7 Quantum Effects in Biology

Until relatively recently it was thought that the strange behaviour of entities reported from quantum physics, was manifested chiefly at the submicroscopic level. However, over the last few years, the role of quantum behaviour in more everyday, macroscopic biological processes has been largely investigated. These studies gave life to *quantum biology*. Quantum biology is an interdisciplinary field investigating quantum effects, such as long lived quantum coherence, that play a role, at the molecular level, within living cells. The field has been supported by a number of experimental discoveries, such as quantum coherence in photosynthesis, quantum tunnelling in enzyme action and olfaction, possible quantum entanglement in avian magnetoreception, bioluminescence, vision and several other areas [39]. All these studies assert that efficiency in nature's biomolecular processes is not wholly explained by conventional theory [10, 41, 42, 45, 46, 49, 105, 106,

107, 108, 109, 110, 134, 147, 156, 176, 198, 212]. In particular, recent experimental evidences suggest that a variety of organisms may harness some of the unique features of quantum mechanics to gain a biological advantage [126].

The biochemical network of the reactions governing the dynamics of biological systems consists of complex molecules structured at the nanoscale and sub-nanoscale. At this scale, the dynamics obey the laws of quantum mechanics. In particular, quantum mechanics and quantum coherence play a crucial role in chemistry. Quantum coherence and entanglement determine the valence structure of atoms and the form of covalent bonds, whereas quantum mechanics fixes the set of allowed chemical compounds and sets the parameters of chemical reactions [135]. Indeed, discrete nature of quantum mechanics ensures that there are only a countable, discrete set of possible chemical compounds. Nevertheless, for many years, the importance of quantum coherence and entanglement for biological processes were not recognized. The reason for this is that biomolecules can contain many atoms (billions in the case of DNA). As molecules become larger and more complex, quantum coherence becomes harder to maintain. Vibrational modes and interactions with the environment decohere quantum superposition, and consequently, most biomolecular mechanisms have traditionally been modelled as essentially classical processes. Recently, however, the central role of quantum coherence has emerged with strong evidence, in particular, in important processes such as photosynthesis, magneto-reception, bioluminescence and sense of smell.

The photosynthesis Plants are able to harvest as much as 95% of the sunlight they soak up, instantly converting the solar energy into chemical energy in 1 million billionth of a second. In this process, plants are employing the basic principles of quantum mechanics to transfer energy from chromophore (photosynthetic molecule) to chromophore until it reaches the so-called Reaction Centre (RC) where photosynthesis, as it is classically defined, takes place [184, 200, 212]. In photosynthesis, the energy of the photons is in small part converted into heat in the form of molecular vibrations, and in large part it captured as an excitons. An exciton is a bound electron-hole residing in the chromophore. Energy transfer occurs via the dipolar interaction between chromophores: the so-called Förster coupling between induced or transition dipoles de-excites one chromophore and excites a neighbouring one [122, 135]. The exciton must propagate through hundreds or thousands of chromophores in order to reach the RC, where charge separation takes place. The process of exciton transport is governed by the laws of quantum mechanics, as quantum mechanics determines the coupling between chromophores and the energy transition rates among them [70, 71].

Fleming et al [71] used femtosecond two-dimensional spectroscopy based on four-wave mixing to demonstrate quantum coherence in photosynthetic transport in bacteria. The result of the experiment was commented by Loyd [135] with these words: "the spectroscopic signature exhibited clear evidence of quantum 'beating' at 77K, a result that was later confirmed at room temperature. The evidence for quantum coherence in photosynthetic transport was essentially incontrovertible"[3]

Fleming et al. [71] proposed that the proof of quantum coherence in photosynthetic energy transport meant that bacteria or plants were performing a *quantum algorithm*[4] (see [103, 157] for a mathematical formalization). Later Lloyd [135] showed that bacteria or plants where performing a particular type of quantum algorithm known as *quantum walk*. In order to explain what quantum walk is, we compare it with the classical random walk. A random walk is a transport mechanism in which the walker takes steps in random directions through the network or structure to be traversed. In a classical random walk, the position of the walker gradually diffuses throughout the network. Because the walker is equally likely to move in any direction, after time t the walker has moved from its original position by a distance proportional to $sqrt t$. In a quantum walk, by contrast, the walker takes a quantum superposition of paths through the network [135]. These paths can exhibit quantum interferences. If the interference is destructive, the walker becomes stuck or 'localized' within the network. On the opposite, if the interference is constructive, the walker moves from its original position by a distance proportional to t. The coherence demonstrated by Fleming showed that - at least over short distances - the excitons were performing a quantum walk. Quantum coherence plays an important role in quantum walk. Lloyd and collaborators [135] build a model of the excitonic transport through the Fenna-Matthews-Olsen (FMO) complex[5], including also the interactions with the environment. The results of the experiment are summarized in the following way by Lloyd [135]:

---

[3]*Quantum beating* is a periodic transition from one state to the other. The double well potential is of ubiquitous use in theoretical physics. A theoretical explanation of the quantum beating phenomenon was obtained modelling the two non-bonding electrons of nitrogen as a quantum particle in a double well potential. For particular values of the parameters of this potential, the ground state and the first excited state have very close energies, forming an almost single degenerate energy level. A superposition of these two states evolves concentrating periodically inside one well or the other, with a frequency proportional to the energy difference [44].

[4]In Chapter 4, we will present the most popular quantum algorithms.

[5]FMO is a water-soluble complex that is present in green sulfur bacteria. It mediates the excitation energy transfer from light-harvesting chlorosomes to the membrane-embedded bacterial reaction centre.

"At low temperatures, the interference between paths in the excitonic walk was predominantly destructive, localizing the exciton to within a few sites of its initial position. At higher temperatures, however, the interaction with the environment decohered the quantum walk sufficiently to remove that destructive interference, allowing the exciton to diffuse throughout the FMO complex towards the reaction centre. At very high temperatures, the decoherence induced by the environment was so destructive that it froze the exciton in place, reducing the efficiency of the transport. The highest efficiency for transport was 290K, the average temperature of the water in which the bacteria lived. Moreover, the transport was robust with respect to variation about this optimal temperature, exhibiting almost 100% efficiency for a range of many tens of degrees K in both directions."

The Lloyd's model not only gave a clear proof of the role of quantum coherence in energy transfer in photosynthetic complexes, but gave also an estimate of the efficiency and of the parameters influencing efficiency of the excitonic dynamics. It is worth to say that recently, the energy transport in photosynthetic complexes has been analyzed through different models with different measures of efficiency. Manzano [147] reports a comprehensive literature about efficiency measures. Manzano mention studies using a Markovian Redfield equation and a generalized BlochRedfield equation. These models used as measure of efficiency the average time that a single excitation spends in the network before being absorbed by the sink. Manzano notices that the Redfield approach correctly describes the dynamics of the system, but it fails to determine the optimal dephasing ratio that minimizes the trapping time. Moreover, this approach gives the unphysical results of a zero trapping time in the limit of strong dephasing (decoherence). In other models reviewed in [147], the efficiency is quantified by the population of the sink in the long time limit. Finally, in other models that do not include a sink, in such a way that the only incoherent dynamics was due to the presence of a dephasing environment, the efficiency is the highest probability of finding the excitation in the outgoing qubit in a time interval $[0, \tau]$, with τ being related to the estimate of the duration of the excitation transfer in real systems.

The avian compass. Some birds sense the earth's magnetic field through mechanisms that cannot be explained by classical mechanics [135]. These mechanisms (a) allow the birds to identify the angle the magnetic field lines make with the earth; (b) are only activated in blue or green light; and (c) are disrupted by oscillating fields a bit above 1 MHz, at the electron's Larmor frequency in the earth's magnetic field. In accordance to the recently proposed explanations for these mechanisms, the absorption of blue or green light creates a radical pair of electrons, separated within an oriented molecule inside the bird's eye. The pair undergoes coherent quantum oscillations between entangled singlet and triplet states, at a rate depending on the orientation of the host molecule with respect to the earth's magnetic field. A spin-dependent electron transition then allows current to flow only if the spins in the pair are in a particular symmetry state. The rate of current flow depends on the molecule's relative orientation to the field. This quantum mechanical mechanism explains all main observed features of the birds' behaviour.

The bioluminescence. A recent finding by Northwestern University's Prem Kumar created quantum entanglement from a biological system [210]. In this study, Kumar's team used green fluorescent proteins, which are responsible for bioluminescence and commonly used in biomedical research. The team tried to entangle the photons generated from the fluorescing molecules within the algae's barrel-shaped protein structure by exposing them to spontaneous four-wave mixing, which is a process in which multiple wavelengths interact with one another to generate new wavelengths.

The Kumar's experiment demonstrated a type of entanglement, called *polarization entanglement*, between photon pairs. Polarization is the orientation of oscillations in light waves. A wave can oscillate vertically, horizontally, or at different angles. In Kumar's entangled pairs, the photons' polarizations are entangled. This means that the oscillation directions of light waves are linked. Most importantly, Kumar also noticed that the barrel-shaped structure surrounding the fluorescing molecules prevented the disruption of the entanglement. Here in the following, we report the words of Kumar summarizing the result of the experiment

"When I measured the vertical polarization of one particle, we knew it would be the same in the other."

"If we measured the horizontal polarization of one particle, we could predict the horizontal polarization in the other particle. We created an entangled state that correlated in all possibilities simultaneously."

The sense of smell. The conventional mechanism postulated for smell is a lock and key model, in which different types of odorant molecule bind to different types of olfactory sensors. The different types of olfactory sensors in the human nose have been identified. In this model, the particular smell of an odorant molecule depends only on its affinity to different sensors. There are several issues with this model. First of all, the smell of a molecule depends at most weakly on shape, the primary determinant of the sensors to which the molecule binds in a lock and key model. Second, smell correlates well with the vibrational spectrum of the molecule: for example, a sulphur-less molecule whose vibrational spectrum exhibits a resonance at the same frequency as the sulphur-hydrogen stretch mode, smells of sulphur. Lloyd comments on this saying: "The nose is evidently a vibrational spectrometer", and reiterates that the only known mechanism that can explain this vibrational sensitivity of the sense of smell is quantum mechanical. The mechanism relies on inelastic tunnelling. Once an odorant molecule docks into the olfactory sensor, electrons can pass through the molecule, and current can flow, only by emitting a phonon of specific frequency to one of the molecule's dominant vibrational modes. The strongest confirming evidence for this mechanism is the ability of fruit flies (drosophila) to smell the difference between an organic molecule and a deuterated version of the same molecule. A molecule whose hydrogens have been replaced by deuterium should bind to the same receptors as the original molecule, with similar affinities. So if the shapebased lock and key mechanism were true, it should smell the same. By contrast, deuterium-carbon bonds vibrate at a rate *sqrt2* slower than hydrogen-carbon bonds: deuteration significantly changes the vibrational spectrum of the molecule. Works to validate further this purely quantum mechanical mechanism are in progress and include X-ray crystallography to identify the precise geometry of the receptor molecules, and creating detailed quantum mechanical models to test and elucidate the specific mechanisms of inelastic tunneling.

Problems

Problem 1.1. Consider a diatomic molecule (consisting of atom 1 and atom 2) in a harmonic potential. If K is the force constant, $\mu = \frac{m_1 m_2}{m_1 + m_2}$ is the reduced mass of the molecule, and x the deviation from the equilibrium position

a) write down the Hamiltonian governing the vibrational motions

b) calculate the ground-state energy and write the wave function associated with the Hamiltonian.

$$\left[\text{Answer: a) } H = \frac{1}{2}\frac{\hbar^2}{\mu}\frac{d^2}{dx^2} + \frac{1}{2}Kx^2. \text{ b) } \exp\left(-\frac{\mu}{2\hbar}\sqrt{\frac{K}{\mu}}x^2\right) \right]$$

Problem 1.2. Write the Hamiltonian of a quantum one dimensional harmonic oscillator (Eq. (1.18)) using the operators O (lowering operator) and O^+ (rising operator), defined as in the following:

$$O = \sqrt{\frac{m\omega}{2\hbar}}\left(x + i\frac{p}{m\omega}\right)$$

$$O^+ = \sqrt{\frac{m\omega}{2\hbar}}\left(x - i\frac{p}{m\omega}\right).$$

where $p = \sqrt{2mE}$ is the momentum of the particle of mass m. Then use the new expression of the Hamiltonian to show that the eigenfunctions of H must also be eigenfunctions of O^+O. *Hint:* to prove that the eigenfunctions of H must also be eigenfunctions of O^+O, use the commutation relation $[O, O^+] \equiv OO^+ - O^+O = 1$.

$$\left[\text{Answer: The Hamiltonian is } H = \hbar\omega\left(O^+O + \tfrac{1}{2}\right)\right]$$

Problem 1.3. Interference occurs when the wave associated to a quantum object may reach the detector along at least two different but intrinsically indistinguishable paths either in real-space or in some configuration space (e.g., potential curves) [16]. The Mach-Zehnder interferometer (Figure 1.2) is often used to demonstrate delocalization and interference for photons over macroscopic distances. It also "visualizes" an idea of coherent energy transport in organic molecules. Suppose each photon lives in a two dimensional Hilbert space spanned by the basis states $|T\rangle$, $|R\rangle$ corresponding to the transmitted and reflected beam. When a photon in the state $|T\rangle$ (see Figure 1.2) hits the beam splitter the electrons of the material absorb it and re-emit it in the new state

$$|\psi\rangle = \frac{1}{\sqrt{2}}|T\rangle + \frac{1}{\sqrt{2}}|R\rangle$$

which is a superposition of the transmitted and reflected states. Assuming that both the beam splitters are balanced, write down the transition matrices BS_1, and BS_2 associated respectively to the beam splitter 1 and to the bean splitter 2 in the basis $\{|T\rangle, |R\rangle\}$.

Hints: describe a photon through a waive function. A photon can live in either in two beams, i.e. the beam in the arm 1 and the beam in the arm 2. Introduce α and β, two complex numbers quantifying the probabilities amplitude to find the arm 1 and in the arm 2, respectively. The probability to find the photon in either of the two arms can be described by a column vector $\begin{pmatrix} \alpha \\ \beta \end{pmatrix}$. Figure 1.3 suggest how to define the transition matrix for the beam splitter.

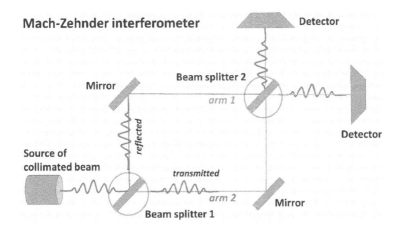

Figure 1.2: The Mach-Zehnder interferometer uses two separate beam splitters to split and recombine the beams of collimate light source, and has two outputs, which are sent to two photodetectors. A beam splitter is a semi transparent mirror which separates a beam of light in two equal intensity beams. The optical path lengths in the two arms may be nearly identical (as in the figure), or may be different (e.g. with an extra delay line). The distribution of optical powers at the two outputs depends on the difference in optical arm lengths and on the wavelength (or optical frequency).

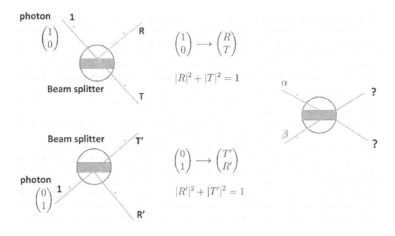

Figure 1.3: A photon can hit the beam splitter coming from arm 1 or coming from arm 2 of the interferometer. Let $(1,0)^T$ and $(0,1)^T$ denote the states of the photon coming from arm 1 and arm 2, respectively. The squared norms of the states must sum to 1, as they describe probabilities. The superposition of the two states is described by the column vector $(\alpha, \beta)^T$.

The beam splitter can receive a photon from the upper or from the lower arm (i.e. arm 1 or arm 2). If $\begin{pmatrix} 1 \\ 0 \end{pmatrix}$ and $\begin{pmatrix} 0 \\ 1 \end{pmatrix}$ denote the states of the photon coming from arm 1 and arm 2, respectively, the beam splitter transforms the beam in the following way respectively:

$$\begin{pmatrix} 1 \\ 0 \end{pmatrix} \longrightarrow \begin{pmatrix} R \\ T \end{pmatrix}$$

$$\begin{pmatrix} 1 \\ 0 \end{pmatrix} \longrightarrow \begin{pmatrix} T' \\ R' \end{pmatrix}$$

with $|R|^2 + |T|^2 = 1$, and $|R'|^2 + |T'|^2 = 1$. Therefore

$$\begin{pmatrix} \alpha \\ \beta \end{pmatrix} = \alpha \begin{pmatrix} 1 \\ 0 \end{pmatrix} + \beta \begin{pmatrix} 1 \\ 0 \end{pmatrix} = \begin{pmatrix} R & T' \\ T & R' \end{pmatrix} \begin{pmatrix} \alpha \\ \beta \end{pmatrix}.$$

The matrix $BS equiv \begin{pmatrix} R & T' \\ T & R' \end{pmatrix}$ is the transition matrix of the beam splitter. For the conservation of probability, this matrix must satisfy the normalization condition

$$|\alpha R + \beta T'|^2 + |\alpha T + \beta R'|^2 = 1.$$

$$\left[\text{Answer: } BS_1 = \frac{1}{\sqrt{2}} \begin{pmatrix} -1 & 1 \\ 1 & 1 \end{pmatrix}; BS_2 = \frac{1}{\sqrt{2}} \begin{pmatrix} 1 & 1 \\ 1 & -1 \end{pmatrix} \right]$$

Problem 1.4.   Consider two states $|0\rangle$ and $|1\rangle$. Suppose that $|0\rangle$ and $|1$ are orthogonal (i.e. $\langle 0|1\rangle = 0$).
    Lets define the state $|\Phi\rangle$ as

$$|\Phi\rangle = \frac{1}{\sqrt{2}}(|0\rangle + e^{i\phi}|1\rangle)$$

where ϕ is constant.

a) Say if $|\Phi\rangle$ is a coherent state.

b) Write the density matrix $\rho = |\Phi\rangle\langle\Phi|$. Say if this matrix describes a coherent state.

c) Write the density matrix ρ' in the case in which ϕ is random, i.e. the mean expectation values of $e^{i\phi}$ is zero. Does ρ' describe a coherent state?

d) Write the probabilities to find the particle in the $|\psi_1\rangle$ state or the $|\psi_2\rangle$ state.

Answers:

a) Yes, $|\Phi\rangle$ is a coherent state because ϕ is constant.

b)

$$\rho = \frac{1}{2}\begin{pmatrix} 1 & e^{i\phi} \\ e^{-i\phi} & 1 \end{pmatrix}$$

ρ describes a coherent state. In fact $\rho^2 = \rho$ and ρ is a projector onto the coherent state $|\Phi\rangle$.

c)

$$\rho' = \frac{1}{2}\begin{pmatrix} 1 & 0 \\ 0 & 1 \end{pmatrix}$$

ρ' does not describes a coherent state (the off-diagonal elements are null).

d) The probabilities to find the particle in the $|\psi_1\rangle$ state or the $|\psi_2\rangle$ state are both equal to 0.5.

Chapter 2

Statistical Physics in Biology

The application domains of statistical physics in biology are multiple. The successful application of statistical physics ideas in biology led to the insertion of statistical physics into the teaching programs of many courses at the university level. They include bioinformatic methods for extracting information content of DNA; DNA sequence comparison, and phylogenetic trees; physical interactions responsible for structure of biopolymers; DNA double helix, secondary structure of RNA, and protein folding. All these fields now named are the oldest and most traditional application domains of statistical physics in biology, for which there is much good literature. So in this chapter, we are instead interested in exposing the most recent application domains, which concern the biology of networks. Indeed, statistical physics concepts are applied also in modelling the complexity of collective behaviour of biological elements, cellular networks, neural networks, and evolution.

2.1 Why Statistical Physics in Biology?

Leonid Mirny, professor of Health Sciences and Technology and Physics at Massachusetts Institute of Technology, explains with these words the most recent opinion - shared by the majority of the actual scientific community - about the use of statistical physics in biology [154, 155, 194].

"There is no unified statistical theory of biology, and I doubt there ever will be any. Nevertheless, I believe that statistical physics is very helpful in thinking about biological phenomena, primarily because biological phenomena are A) stochastic, so randomness plays a huge role and B) there are lots of players in biological systems. These players can be many molecules in a body, or can be many cells in a human body, or can be many organisms in a population. Again, the two underlying things are stochasticity and the large number of players."

"The field of biology that really lives above all the complexity of molecular interactions is genetics. Genetics is interested in the emergence of new sorts of, say, mutations and then how these mutations spread in the population. For example, you have a population of white mice and then one of them becomes grey, and now you have grey mice which are able to better hide within their environment. So, the grey would spread in the population. Thats an evolutionary process and the field of biology that deals with this is genetics, or population genetics. So population genetics is generally interested in this kind of collective phenomena."

"If you look at certain processes, like a developmental process: you had one cell, becomes two cells, four cells, then it acquires a sort of shape, and these cells form body parts. So, it's a very reliable process, despite some underlined stochasticity. So statistical physics is certainly interested in this process. It may be reminiscent of some of the features, for example, of condensation of water, where initially very disordered system, each molecule is still sort of fluctuating and jiggling independently, though it will all end up being a snowflake, a very regular, very structured beautiful organization."

In these thoughts of Mirny, we find some recurring concepts such as: stochasticity, complex biological networks, temporal evolution. The need for computational models of dynamics of complex biologic networks for the purpose of a better understanding of the processes underlying them and the saving of time and money required in expensive laboratory experiments have pushed more and more research into the development of simulation algorithms [38, 130]. For such purposes, the computational models and the simulation algorithms, have to be necessarily able to mimic the stochasticity inherent in various phenomena biologists in an efficient way even for biological systems of large dimensions, of stiff and/or non-linear dynamics, and represented by dense interaction graphs. The well-established stochastic simulation algorithm, as well as the recent ones, are based on concepts of statistical physics, such as master equation, Markov processes, and statistical physics formulation of thermodynamics. In the next section, we will present these concepts, that have been dealt with extensively in Lecca [127] and Lecca et al.

[130, 131]. Lecca [127] and Lecca et al. [130, 131] are the main sources of the contents of the next sections.

2.2 Markov Processes

A Markov process is a special case of a stochastic process. Markov processes are used to model randomness, since it is much more tractable than a general stochastic process. A general stochastic process is a random function $f(X;t)$, where X is a stochastic variable and t is time. The definition of a stochastic variable consists in specifying (i) a set of possible values (called "set of states" or "sample space") and (ii) a probability distribution over this set. For instance, the set of states may be *discrete*, i.e. the number of molecules of a certain component in a reacting mixture. Alternatively, the set may be *continuous* in a given interval, e.g. one velocity component of a Brownian particle and the kinetic energy of that particle. The set may be partly discrete and partly continuous, e.g. the energy of an electron in the presence of binding centres. Moreover the set of states may be *multidimensional*: in this case X is represented by a vector $\vec{X}$. Examples: $\vec{X}$ may stand for the three velocity components of a Brownian particle or for the collection of all numbers of molecules of the various components in a reacting mixture.

The probability distribution, on a continuous one-dimensional domain Ω_X, is given by a function $P(x)$ that is non-negative

$$P(x) \geq 0$$

and normalized in the sense

$$\int_{\Omega_X} P(x)dx = 1.$$

The probability that X has a value between x and $x + dx$ is

$$P(x)dx$$

Often in physical sciences a probability distribution is visualized by an "ensemble". Adopting an ensemble-like representation, a fictitious set of an arbitrary large number $\mathcal{N}$ of quantities, all having different values in the given range, is introduced. In such a way the number of these quantities having a value between x and $x + dx$ is $\mathcal{N}P(x)dx$. Therefore, the probability distribution is replaced with a density distribution of a large number of "samples". This does not affect any simulation result, since it is a mere "linguistic" convenience in talking about probabilities. Nevertheless, it is worth to say that it could be a suitable representation of real systems. For instance, a biochemical system consisting of a large number of identical replica constitutes to a certain extent a physical realization of an ensemble. The molecules of an ideal gas may serve as an ensemble representing the Maxwell probability distribution for the velocity.

Finally, we note that if we include in $P(x)$ delta functions, i.e.

$$P(x) = \sum_n p_n \delta(x - x_n) + \tilde{P}(x),$$

where $\tilde{P}$ is finite or at least integrable and non-negative, $p_n > 0$, and

$$\sum_n p_n + \int \tilde{P}(x)dx = 1$$

we indeed are representing a set of discrete states x_n with probability p_n embedded in a continuous range. If $P(x)$ consists of δ functions alone (i. e. $\tilde{P}(x) = 0$, then it can also be considered as a probability distribution p_n on the discrete set of states x_n.

In general, a stochastic process defines the joint probability densities for values $x_1, x_2, x_3, \ldots$ at times $t_1, t_2, t_3, \ldots$ respectively

$$p(x_1, t_1; x_2, t_2; x_3, t_3; \ldots). \tag{2.1}$$

If all such probabilities are known, the stochastic process is fully specified. Using (2.1) the conditional probabilities can be defined as usual

$$p(x_1, t_1; x_2, t_2; \ldots | y_1, \tau_1; y_2, \tau_2; \ldots) = \frac{p(x_1, t_1; x_2, t_2; \ldots | y_1, \tau_1; y_2, \tau_2; \ldots)}{p(y_1, \tau_1; y_2, \tau_2; \ldots)}$$

where $x_1, x_2, \ldots$ and $y_1, y_2, \ldots$ are values at times $t_1 \geq t_2 \geq \cdots \geq \tau_1 \geq \tau_2 \geq \ldots$. We can see now that Markov process has a very attractive property: it has no memory. For a Markov process

$$p(x_1, t_1; x_2, t_2; \ldots | y_1, \tau 1; y_2, \tau_2; \ldots) = p(x_1, t_1; x_2, t_2; \ldots | y_1, \tau_1)$$

the probability to reach a state x_1 at time t_1 and state x_2 at time t_2, if the state is y_1 at time τ_1, is independent of any previous state, with times ordered as before. This property makes it possible to construct any of the probabilities (2.1) by a *transition probability* $p_\to(x, t | y, \tau)$, $(t \geq \tau)$, and an initial probability distribution $p(x_n, t_n)$:

$$p(x_1, t_1; x_2, t_2; \ldots x_n, t_n) =$$
$$p_\to(x_1, t_1 | x_2, t_2) p_\to(x_2, t_2 | x_3, t_3) \ldots p_\to(x_{n-1} t_{n-1} | x_n, t_n) p(x_n, t_n) \tag{2.2}$$

A consequence of the Markov property is the Chapman-Kolmogorov equation

$$p_\to(x_1, t_1 | x_3, t_3) = \int p_\to(x_1, t_1 | x_2, t_2) p_\to(x_2, t_2 | x_3, t_3) dx_2. \tag{2.3}$$

2.3 The Chemical Master Equation

The master equation is a differential form of the Chapman-Kolmogorov equation (2.3). In this book we will use the expression master equation only for jump processes. Jump processes are characterized by discontinuous changes, that is there is a bounded and non-vanishing transition probability per unit time

$$w(x|y,t) = \lim_{\Delta t \to 0} \frac{p_{\to}(x, t + \Delta t|y, t)}{\Delta t}$$

for some y such that $|x - y| > \epsilon$. Here, the function $w(x|y;t) = w(x|y)$.

The master equation for jump processes is

$$\frac{\partial p(x,t)}{\partial t} = \int \Big(w(x|x')p(x', t) - w(x'|x)p(x, t) \Big) dx'. \tag{2.4}$$

The first part of the integral in Eq. (2.4) is the *gain of probability* from the state x' and the second part is the *loss of probability* to x'. The solution is a probability distribution for the state space. Analytical solutions of the master equation are possible to calculate only for simple special cases.

A reaction R is defined as a jump to the state $\vec{X}$ to a stare $\vec{X}_R$, where $\vec{X}, \vec{X}_R \in \mathbb{Z}_+^N$. The propensity $w(\vec{X}_R) = \tilde{v}(\vec{X})$ is the probability for transition from $\vec{X}_R$ to $\vec{X}$ per unit time. A reaction can be written as

$$\vec{X}_R \xrightarrow{w(\vec{X}_R)} \vec{X}$$

The difference in molecules numbers $\vec{n}_R = \vec{X}_R - \vec{X}$ is used to write the master equation (2.4) for a system with M reactions

$$\frac{dp(\vec{X},t)}{dt} = \sum_{i=1}^{M} w(\vec{X} + n)p(\vec{X} + \vec{n}_R, t) - \sum_{i=1}^{M} w(\vec{X})p(\vec{X}, t) \tag{2.5}$$

This special case of master equations is called the *chemical master equation* (CME) [150, 213]. It is fairly easy to write: however, solving it is quite another matter. The number of problems for which the CME can be solved analytically is even fewer than the number of problems for which the deterministic reaction-rate equations can be solved analytically. Attempts to use master equation to construct tractable time-evolution equations are also usually unsuccessful, unless all the reactions in the system are simple monomolecular reactions [88]. Let us consider for instance a deterministic model of two metabolites coupled by a bimolecular reaction (Jöberg [116]), as shown in Fig. 2.1. The set of differential equation describing the dynamic of this model is given in Table 2.1, where the $[A]$ and $[B]$ are the concentrations of metabolite

A and metabolite B, while k, K, and μ determine the maximal rate of synthesis, the strength of the feedback, and the rate of degradation, respectively.

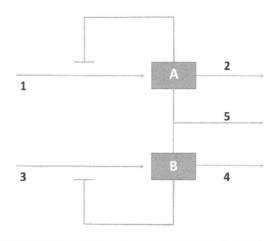

Figure 2.1: Two metabolites A and B coupled by a bimolecular reactions. Adapted from [116].

Table 2.1: Reactions of the chemical model displayed in Figure 2.1. "Nr" in this table corresponds to the number on the arrows of the network in Figure 2.1. Adapted from Lecca et al. [131].

Nr.	Reaction	Rate equation	Process
1	$\emptyset \xrightarrow{v_1([A])} A$	$v_1([A]) = \frac{k_1}{1+[A]K_1}$	synthesis
2	$A \xrightarrow{v_2([A])} \emptyset$	$v_2([A]) = \mu[A]$	degradation
3	$\emptyset \xleftarrow{v_3([B])} B$	$v_3([B]) = \frac{k_2}{1+[B]/K_2}$	synthesis
4	$B \xrightarrow{v_4([B])} \emptyset$	$v_4([B]) = \mu[B]$	degradation
5	$A + B \xrightarrow{v_5([A],[B])} \emptyset$	$v_5([A],[B]) = k_3[A][B]$	bimolecular reaction

In the formalism of the Markov process, the reactions in Table 2.1 are written as in Table 2.2. The CME equation for the system of two metabolites in Fig. 2.1 looks fairly complex as in Table 2.3.

Table 2.2: Reactions of the chemical model depicted in Fig. 2.1, their propensity and corresponding "jump" of state vector $\vec{n}_R^T$. V is the volumes in which the reactions occur. Adapted from Lecca et al. [131].

Nr.	Reaction	$w(\vec{x})$	$\vec{n}_R^T$
1	$\emptyset \xrightarrow{w_1(a)} A$	$w_1(a) = Vk_1/(1 + a/VK_1))$	$(-1, 0)$
2	$A \xrightarrow{w_2(a)} \emptyset$	$w_2(a) = \mu a$	$(1, 0)$
3	$\emptyset \xrightarrow{w_3(b)} B$	$w_3(b) = VK_2/(1 + b/(VK_2))$	$(0, -1)$
4	$B \xrightarrow{w_4(b)} \emptyset$	$w_4(b) = \mu b$	$(0, 1)$
5	$A + B \xrightarrow{w_5(a,b)} \emptyset$	$w_5(a, b) = k_2 ab/V$	$(1, 1)$

Table 2.3: Set of chemical master equations describing the metabolites interaction showed in Fig. 2.1. Adapted from Lecca et al. [131].

$$\frac{\partial(0,0,t)}{\partial t} = \mu p(1,0,t) + \mu p(0,1,t) + \frac{k_3}{V} p(1,1,t) - V(k_1 + k_2)p(0,0,t)$$

$$\frac{\partial(0,b,t)}{\partial t} = V\frac{k_2}{1 + \frac{b-1}{VK_2}} p(0,b-1,t) +$$

$$+ \mu p(1,b,t) + \mu(b+1)p(0,b+1,t) + \frac{k_3}{V}(b+1)p(1,b+1,t) -$$

$$- \left(V\left(k_1 + \frac{k_2}{1 + \frac{b}{VK_2}} \right) + \mu b \right) p(0,b,t)$$

$$\frac{\partial p(a,0,t)}{\partial t} = V\frac{k_1}{1 + \frac{a-1}{VK_1}} p(a-1,0,t) +$$

$$+ \mu(a+1)p(a+1,0,t) + \mu p(a,1,t) +$$

$$+ \frac{k_3}{V}(a+1)p(a+1,1,t) -$$

$$- \left(V\left(\frac{k_1}{1 + \frac{a}{VK_1}} + k_2 \right) + \mu a \right) p(a,0,t)$$

$$\frac{\partial p(a,b,t)}{\partial t} = V\frac{k_1}{1 + \frac{a-1}{VK_1}} p(a-1,b,t) + V\frac{k_2}{1 + \frac{b-1}{VK_2}} p(a,b-1,t) +$$

$$+ \mu(a+1)p(a+1,b,t) + \mu(b+1)p(a,b+1,t) +$$

$$+ \frac{k_3}{V}(a+1)(b+1)p(a+1,b+1,t) -$$

$$- \left(V\left(\frac{k_1}{1 + \frac{a}{VK_1}} + \frac{k_2}{1 + \frac{b}{VK_2}} \right) + \mu(a+b) + \frac{k_3}{V}ab \right) p(a,b,t)$$

2.4 Chemical Master Equation and Curse of Dimensionality

Chemical master equation is fairly easy to write: however, solving it is quite another matter. The number of problems for which the CME can be solved analytically is even fewer than the number of problems for which the deterministic reaction-rate equations can be solved analytically. Attempts to use master equation to construct tractable time-evolution equations are also usually unsuccessful, unless all the reactions in the system are simple monomolecular reactions [88].

Analytical solutions of the master equation are possible to calculate only for simple special cases. Despite the promising insights it offers, the CME suffers from the *curse of dimensionality* [99] as the size of the state space grows exponentially with the number of species involved.

The curse of dimensionality, a term initially introduced by Richard Bellman [25] refers to the problem of finding models for data embedded in a highly dimensional space. The more features we have, the more data points we need in order to fill space. From a statistical perspective, curse of dimensionality relates to the fact that the convergence of any estimator to the true value of a smooth function defined on a space of high dimension is very slow. Stochastic biochemical systems can be modelled by the chemical master equations, but the major challenge posed by these models, is the curse of dimensionality which occurs when one attempts to integrate these equations. The complexity of the CME is due to the fact that it involves one state space dimension for each species involved in the reactions. This means that classic techniques such as finite elements involve a number of degrees of freedom growing exponentially with the number of reacting species. Ammar et al. [13] estimated that finite element-based techniques are limited by this reason to problems on the order of 20 state space dimensions, by using sparse grid techniques. The problem of curve of dimensionality can be better described from a geometrical point of view. Consider a hypercube with side length equal to 1, in an n-dimensional space. The volume of the hypercube is 1. If we want to allocate that volume among N smaller cubes (each containing a data point) uniformly distributed in the n-dimensional hypercube, each small cube will have a volume equal to $1/N$. Their side length d is

$$d = \left(\frac{1}{N}\right)^{\frac{1}{n}}$$

For a finite N, d converges to 1 when n goes to infinity. That is, the new smaller cubes each have almost the same volume as the bigger cube. In 1-infinite dimensional space we can put N cubes of volume 1 inside a cube of volume 1. Keeping the volume of the smaller cubes fixed requires an exponentially growing number of data points N, when the dimension increases.

Sometimes, instead of using CME, researchers use the indirect approach of simulating trajectories by the stochastic simulation algorithm and variants such as τ-leaping. Unfortunately, these approaches require a large number of trajectories when it comes to finding the probability distribution of the underlying Markov process. The finite state projection (FSP) is an alternative approach that seeks to approximate directly this distribution by restricting the CME to a more tractable size [65, 95, 100, 141, 142, 160, 162, 204]. This method has recently been integrated into a larger framework that utilizes novel single-cell experimental methods to successfully identify predictive models of several gene regulatory networks [228], as well as to compare different CME models of single-cell, or single-molecules responses [83]. Unfortunately, more complex biological systems require a large state space that is beyond the reach of the standard FSP. This challenge motivated many developments, e.g., Krylov-FSP [66, 219], multiple-time-interval FSP [161], wavelet [209], quantized tensor train decomposition [120, 220, 221], SSA-driven FSP [65, 142] and many others. Although these methods are promising for the purpose of solving the CME, the most widely used tool is the Gillespie stochastic simulation algorithm and its most current and most efficient variants [90, 91].

2.5 Discrete Approach to Chemical Kinetics

The solution of the set of ordinary differential equations (ODE) in a deterministic model, describes the time-evolution of the system, i.e. the changes in time of the state vectors $\vec{x}$ of the system. If the system includes N molecular species X_i $(i = 1, \ldots, N)$, an ODE model is a set of equations of this form

$$\frac{d[X_i]}{dt} = f(X_1, X_2, \ldots X_N)$$

where $[X_i]$ is the concentration of species i and $F : \mathbb{R}^+ \to \mathbb{R}^+$ is a deterministic function describing the behaviour of the rate of change of $[X_i]$. The solution of these equations is not the precise number of reactive collisions, but the average. The actual number fluctuates around it and in order to find the resulting fluctuation in x_j around the macroscopic values determined by these solutions we need to switch to a molecular approach to the chemical kinetics.

Switching from a deterministic model to a stochastic one requires to switch the unit of measure from concentration to number of molecules. In the stochastic model, this is an integer representing the number of molecules of the species, but in the deterministic model, it is a concentration, measured in M (moles per litre). Then for a concentration of X of $[X]$ M in a volume of V litres, there are $[X]V$ moles of X and hence $n_A[X]V$ molecules, where $n_A \simeq 6.023 \times 10^{23}$ is the Avogadro's constant (the number of molecules in a mole). The second issue that needs to be addressed is the rate constant conversion. Much of the literature on biochemical reaction is dominated by a continuous deterministic modelling of kinetics. Indeed, the rate constants are estimated by fitting the experimental data with deterministic models. Consequently, where rate constants are documented, they are usually expressed in continuous measurement units. In the following we review the expression of the reaction propensity and the formulae that convert the rate constants from deterministic to stochastic framework.

2.5.1 *Effective Reactions are Inelastic Collisions*

For a reaction to produces products, molecules must collide with sufficient energy to create a transition state. Ludwig Boltzmann developed a very general idea about how energy was distributed among systems consisting of many particles. He said that the number of particles with energy E would be proportional to the value $\exp[-E/k_B T]$. The Boltzmann distribution predicts the distribution function for the fractional number of particles N_i/N occupying a set of states i which each have energy E_i:

$$\frac{N_i}{N} = \frac{g_i e^{-E_i/k_B T}}{Z(T)}$$

where k_B is the Boltzmann constant, T is temperature (assumed to be a sharply well-defined quantity), g_i is the degeneracy, or number of states having energy E_i, N is the total number of particles:

$$N = \sum_i N_i,$$

and $Z(T)$ is called the *partition function*

$$Z(T) = \sum_i g_i e^{-E_i/k_B T}$$

Alternatively, for a single system at a well-defined temperature, it gives the probability that the system is in the specified state. The Boltzmann distribution applies only to particles at a high enough temperature and low enough density that quantum effects can be ignored.

James Clerk Maxwell used Boltzmann's ideas and applied them to the particles of an ideal gas to produce the distribution bearing both men's names (the Maxwell-Boltzmann distribution). Maxwell also used for the energy E the formula for kinetic energy $E = (1/2)mv^2$, where v is the velocity of the particle. The distribution is best shown as a graph which shows how many particles have a particular speed in the gas. It may also be shown with energy rather than speed along the x axis. Two graphs are shown in Figures 2.2 and 2.3.

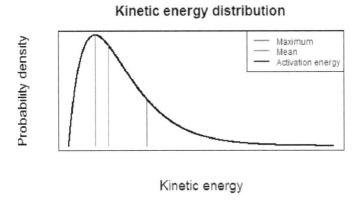

Kinetic energy distribution

Probability density

Kinetic energy

Figure 2.2: Maxwell-Boltzmann distribution of kinetic energy. Since the curve shape is not symmetric, the average kinetic energy will always be greater than the most probable. For the reaction to occur, the particles involved need a minimum amount of energy - the activation energy. This energy is the difference of average energy of reactants and the average at the transition state. Reactions with large activation energies ($\gg k_B T$) are slow.

Consider a bi-molecular reaction of the form

$$S_1 + S_2 \longrightarrow \dots \tag{2.6}$$

the right-hand side is not important in this discussion. This reaction means that a molecule of S_1 is able to react with a molecule of S_2 if the pair happen to collide with one another with sufficient energy, while moving around randomly, driven by Brownian motion. Consider a single pair of such molecules in a closed volume V. It is possible to use statistical mechanics arguments to understand the physical meaning of the *propensity* (i.e. hazard) of molecules colliding. Under the assumptions that the volume is not too large or well stirred, in thermal equilibrium, and constant, it can be rigorously proven that the *collision propensity* (also called collision *hazard, hazard function* or *reaction hazard*) is constant, provided that the temperature is constant. Since the molecules are uniformly distributed throughout the volume and this distribution does not depend on time, then the probability

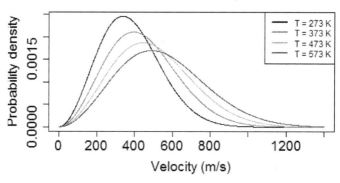

Figure 2.3: **Maxwell-Boltzmann velocity distributions for Argon at different thermodynamic temperatures. As temperature increases, the curve will spread to the right and the value of the most probable kinetic energy will decrease. At temperature increases the probability of finding molecules at higher energy increases. Note also that the area under the curve is constant since total probability must be one.**

that the molecules are within reaction distance is also independent of time [88, 89]. Here we briefly review the theory developed by Gillespie [88, 89] by highlighting the physical basis of the stochastic formulation of chemical kinetics. Consider now that the system composed of a mixture of the two molecular species, S_1 and S_2 in gas-phase and in thermal, but not necessarily chemical equilibrium inside the volume V. Lets assume that the S_1 and S_2 molecules are hard spheres of radii r_1 and r_2, respectively. A collision will occur whenever the centre-to-centre distance between an S_1 molecule and an S_2 molecule is less than $r_{12} = r_1 + r_2$. To calculate the molecular collision rate, we consider an arbitrary 1-2 molecular pair, and we denote by v_{12} the speed of the molecule 1 relative to molecule 2. Then, in the next small time interval δt, molecule 1 will sweep out relative to molecule 2 a collision volume

$$\delta V_{coll} = \pi r_{12}^2 v_{12} \delta t$$

i.e. if the centre of molecule 2 happens to lie inside δV_{coll} at time t, then the two molecules will collide in the time interval $(t, t + \delta t)$. Now, the classical procedure would estimate the number of S_2 molecules whose centres lie inside δV_{coll}, divide the number by δt, and then take the limit $\delta \to 0$ to obtain the rate at which the S_1 molecule is colliding with S_2 molecules. However, this procedure suffers from the following difficulty: as $\delta V_{coll} \to 0$, the number of S_2 molecules whose centres lie inside δV_{coll} will be either 1 or 0, with the latter possibility become more and more likely as the limiting process proceeds. Then, in the limit of vanishingly small δt, the concept of "number of molecules whose centre lies inside δV_{coll}" is physically meaningless.

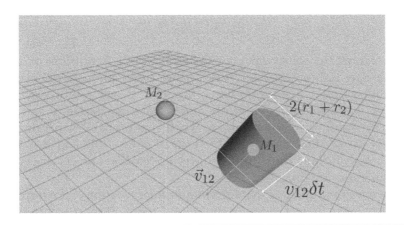

Figure 2.4: A schematic representation of the collision between two molecules, M_1 and M_2 seen from the reference system integral with the particle 2. The collision volume δV_{coll} which molecule 1 will sweep out relative to molecule 2 in the next small time interval δt.

To overcome this difficulty we can exploit the assumption of thermal equilibrium. Since the system is in thermal equilibrium, the molecules will at all times be distributed randomly and uniformly throughout the containing volume V. It follows that, the *probability* that the centre of an arbitrary S_2 molecule will be found inside δV_{coll} at time t will be given by the ratio $\delta V_{coll}/V$; note that this is true even in the limit of vanishingly small δV_{coll}. If we now average this ration over the velocity distributions of S_1 and S_2 molecules, we may conclude that the average probability that a particular 1-2 molecular pair will collide in the next vanishingly small time interval δt is

$$\overline{\delta V_{coll}/V} = \frac{\pi r_{12}^2 \overline{v_{12}}\delta t}{V} \tag{2.7}$$

For Maxwellian velocity distributions the average relative speed $\overline{v_{12}}$ is

$$\overline{v_{12}} = \left(\frac{8kT}{\pi m_{12}}\right)^{\frac{1}{2}}$$

where k is the Boltzmann's constant, T the absolute temperature, and m_{12} the reduced mass $m_1 m_2/(m_1 + m_2)$. If at time t there are X_1 molecules of the species S_1 and X_2 molecules of the species S_2, making a total of $X_1 X_2$ distinct 1-2 molecular pairs, then if follows from (2.7) that the probability that a 1-2 collision will occur somewhere inside V in the next infinitesimal time interval $(t, t + dt)$ is

$$\frac{X_1 X_2 \pi r_{12}^2 \overline{v_{12}} dt}{V} \tag{2.8}$$

We cannot rigorously calculate the number of 1-2 collisions occurring in V in any infinitesimal interval, but we can rigorously calculate the probability of a 1-2 collision occurring in V in any infinitesimal time interval. Consequently, we really ought to characterize a system of thermally equilibrized molecules by a collision probability per unit time (namely the coefficient of dt in (2.8) instead of by a collision rate. This is why these collisions constitute a stochastic Markov process instead of a deterministic rate process.

Then we can conclude that for a bimolecular reaction of the form (2.6), the probability that a randomly chosen A-B pair will react according to R in next dt is

$$
\begin{aligned}
P_{react} &= \left\{ \left(\frac{(\overline{v_{12}} dt)(\pi r_{12}^2)}{V} \right) \times \exp[-E/(k_B T)] \right\} \times X_1 X_2 \\
&= \left\{ \left(\frac{\overline{v_{12}}(\pi r_{12}^2)}{V} \exp(-E/(k_B T) \right) \right\} X_1 X_2 dt
\end{aligned} \tag{2.9}
$$

2.5.2 Factors Affecting Reaction Rate

The reaction rate for a reactant or product in a particular reaction is defined as the amount of the chemical that is formed or removed (in moles or mass units) per unit time per unit volume. The main factors that influence the reaction rate include: the physical state of the reactants, the volume of the reaction chamber, the temperature at which the reaction occurs, and whether or not any catalysts are present in the reaction.

Physical state

When reactants are in the same phase, as in aqueous solution, thermal motion brings them into contact. However, when they are in different phases, the reaction is limited to the interface between the reactants. Reaction can only occur at their area of contact, in the case of a liquid and a gas, at the surface of the liquid. The more finely divided a solid, the greater its surface area per unit volume, and the more contact it makes with the other reactant, thus the faster the reaction.

Volume

The reaction propensity is inversely proportional to the volume. To prove this relationship, consider two molecules, Molecule 1 and Molecule 2. Let the molecules positions in space be X_1 and x_2 respectively. If x_1 and x_2 are uniformly and independently distributed over volume V, for a sub-region of space D with volume V', the probability that a molecule is inside D is

$$\Pr(x_i \in D) = \frac{V'}{V} \quad i = 1,\ 2$$

The probability that Molecule 1 and Molecule 2 are within a reacting distance r of one another at any given instant of time (assuming that r is much smaller than the dimensions of the reaction chamber, so that boundary effects can be ignored), is

$$\Pr(|x_1 - x_2| < r) = \mathbb{E}(\Pr(|x_1 - x_2| < r|x_2)),$$

but the conditional probability will be the same for any x_2 away from the boundary, so that the expectation $\mathbb{E}$ is redundant, and we can state that

$$\mathbb{E}(\Pr(|x_1 - x_2| < r|x_2)) = \Pr(|x_1 - x_2| < r) = \Pr(x_i \in D) = \frac{4\pi r^3}{3V}.$$

Temperature

A molecule has more energy when it is heated As a consequence, the more energy it has, the more chances it has to collide with other reactants. Thus, at a higher temperature, more collisions occur.

The temperature dependency of the reaction rate coefficient k is described by the empirical Arrhenius law:

$$k = A \exp\left[-\frac{E_a}{RT}\right] \tag{2.10}$$

where E_a is the activation energy and R is the gas constant. Since at temperature T the molecules have energies given by the Boltzmann distribution, one can expect the number of collisions with energy greater than E_a to be proportional to $\exp[-E_a/RT]$. A is the frequency factor. The value of this factor is the number of collisions between reactants that have the correct orientation to lead to the products. The values for A and E_a are reaction-specific. Either increasing the temperature or decreasing the activation energy (for example through the use of catalysts) result in an increase in rate of reaction.

Various other expressions alternative to Arrhenius equation are sometimes found to be more accurate in particular situations. One example comes from the "collision theory" of chemical reactions, developed by Max Trautz and William Lewis between 1916-18 [206]. According to this theory, molecules react if they collide with a relative kinetic energy along their line-of-centres that exceeds E_a. This leads to an expression very similar to the Arrhenius equation, with the difference that the pre-exponential factor A is not constant but instead is proportional to the square root of temperature. This reflects the fact that the overall rate of all collisions, reactive or not, is proportional to the average molecular speed which in turn is proportional to $\sqrt{T}$. Another Arrhenius-like expression appears in the Transition State Theory of chemical reactions, formulated by Wigner, Eyring, Polanyi and Evans in the 1930's. This takes various forms, but one of the most common is:

$$k = \frac{k_B T}{h} \exp\left[-\frac{\Delta G}{RT}\right]$$

where ΔG is the Gibbs free energy of activation, k_B is Boltzmann's constant, and h is Planck's constant. In this expression, it is worth to note that the free energy itself is a temperature dependent quantity. The free energy of activation includes an entropy term and an enthalpy term, both depending on temperature. Consequently, after some algebraic manipulations we end up with an expression that again takes the form of an Arrhenius exponential multiplied by a slowly varying function of T.

The precise form of the temperature dependence is reaction-specific, and can be calculated using formulas from statistical mechanics that involve the partition functions of the reactants and those of the activated complex.

Catalysts

A catalyst is a substance that increases the rate of a reaction by providing a different reaction mechanism to occur with a lower activation energy. In autocatalysis a reaction product is itself a catalyst for that reaction possibly leading to a chain reaction. Proteins that act as catalysts in biochemical reactions are called enzymes (a very well-known simple model of enzymatic reactions are given by the Michaelis-Menten kinetics).

The formulation of stochastic chemical kinetics of Gillespie assumes that temperature and volume of container do not change in time. However in the biological context these assumption may not hold and may lead to obtain wrong simulation results. Lecca [128] proposed an extension of the Gillespie algorithm to overcome this limitation.

2.6 Stochastic Simulation Algorithm

In this section we introduce the foundation of the stochastic simulation algorithm of Gillespie. If the system is in the state $\vec{X} = \{X_1, \ldots, X_N\}$ at time t, computing its stochastic evolution means "moving the system forward in time". In order to do that we need to answer two questions.

1. When will the next reaction occur?
2. What kind of reaction will it be?

Due to the essentially random nature of chemical interactions, these two questions are answerable only in a probabilistic way.

Let us introduce the function $P(\tau, \mu)$ defined as the probability that, given the state $\vec{X}$ at time t, the next reaction in the volume V will occur in the infinitesimal time interval $(t+\tau, t+\tau+d\tau)$, and will be an R_μ reaction. $P(\tau, \mu)$ is called *reaction probability density function*, because it is a joint probability density function on the space of the *continuous* variable τ $(0 \leq \tau < \infty)$ and the *discrete* variable μ $(\mu = 1, 2, \ldots, M)$.

The values of the variables τ and μ will give us answer to the two questions mentioned above. $P(\tau, \mu)$ can be written as the product of $P_0(\tau)$, the probability that given the state $\vec{X}$ at time t, no reaction will occur in the time interval $(t, t+dt)$, times $a_\mu d\tau$, the probability that an R_μ reaction will occur in the time interval $(t + \tau, t + \tau + d\tau)$

$$P(\mu, \tau)d\tau = P_0(\tau)a_\mu dt \tag{2.11}$$

In turn $P_0(\tau)$ is given by

$$P_0(\tau' + d\tau') = P_0(\tau')\left[1 - \sum_{i=1}^{M} a_i d\tau'\right] \tag{2.12}$$

where $[1 - \sum_{i=1}^{M} a_i d\tau']$ is the probability that no reaction will occur in time $d\tau'$ from the state $\vec{X}$. Therefore

$$P_0(\tau) = \exp\left[-\sum_{i=1}^{M} a_i \tau\right] \tag{2.13}$$

Inserting (2.12) into (2.11), we find the following expression for the reaction probability density function

$$P(\mu, \tau) = \begin{cases} a_\mu \exp(-a_0\tau) & \text{if } 0 \leq \tau < infty \\ 0 & otherwise \end{cases} \tag{2.14}$$

where a_μ is the reaction propensity for the reaction R_μ expressed as the product of the stochastic rate constant c_μ and the number h_μ of distinct combination of reactant molecules of R_μ

$$a_\mu = c_\mu \cdot h_\mu. \tag{2.15}$$

$$a_0 \equiv \sum_{i=1}^{M} a_i \equiv \sum_{i=1}^{M} h_i c_i \tag{2.16}$$

The expression for $P(\mu, \tau)$ in (2.14) is, like the master equation in (2.5), a rigorous mathematical consequence of the fundamental hypothesis of stochastic chemical kinetics [87, 88, 90]. Notice finally that $P(\tau, \mu)$ depends on all the reaction constants (not just on c_μ) and on the current numbers of all reactant species (not just on the R_μ reactants).

On each step the Direct Method of Gillespie algorithm generates two random numbers r_1 and r_2 from a set of uniformly distributed random numbers in the interval $(0,1)$. The time for the next reaction to occur is given by $t + \tau$, where τ is given by

$$\tau = \frac{1}{a_0} \ln\left(\frac{1}{r_1}\right) \tag{2.17}$$

The index μ of the occurring reaction is given by the smallest integer satisfying

$$\sum_{j=1}^{\mu} a_j > r_2 a_0 \tag{2.18}$$

The system states are updated by $X(t + \tau) = X(t) + v_\mu$, then the simulation proceeds to the next occurring time.

We now summarize the steps of Direct Method of Gillespie simulation algorithm.

1. Initialization: set the initial numbers of molecules for each chemical species; input the desired values for the M reaction constants $c_1, c_2, \ldots, c_M$. Set the simulation time variable t to zero and the duration T of the simulation.

2. Calculate and store the propensity functions a_i for all the reaction channels $(i = 1, \ldots, M)$, and a_0.

3. Generate two random numbers r_1 and r_2 in $Unif(0,1)$.

4. Calculate τ according to (2.17)

5. Search for μ as the smallest integer satisfying (2.18).

6. Update the states of the species to reflect the execution of μ (e.g. if $R_\mu : S_1 + S_2 \to 2S_1$, and there are X_1 molecules of the species S_1 and X_2 molecules of the species S_2, then increase X_1 by 1 and decrease X_2 by 1). Set $t \leftarrow t + \tau$.

7. If $t < T$ then go to step 2, otherwise terminate.

Note that the random pair (τ, μ), where τ is given by (2.17) and μ by (2.18), is generated according to the probability density function in (2.14). A rigorous proof of this fact may be found in [87]. Suffice here to say that (2.17) generates a random number τ according to the probability density function

$$P_1(\tau) = a_0 \exp(-a_0 \tau) \tag{2.19}$$

while (2.18) generates an integer μ according to the probability density function

$$P_2(\mu) = \frac{a_\mu}{a_0} \tag{2.20}$$

and the stated result follows because

$$P(\tau, \mu) = P_1(\tau) \cdot P_2(\mu)$$

To generate random numbers between 0 and 1 we can do as follows. Let $F_X(x)$ be a distribution function of an exponentially distributed variable X and let $U \sim Unif[0,1)$ denote an uniformly distributed random variable U on the interval $[0,1)$.

$$F_X(x) = \begin{cases} 1 - e^{-ax} & \text{if } x \geq 0 \\ 0 & \text{if } x < 0 \end{cases} \tag{2.21}$$

$F_X(x)$ is a continuous non-decreasing function and this implies that it has an inverse F_X^{-1}. Now, let $X(U) = F_X^{-1}(U)$ and we get the following

$$P(X(U) \leq x) = P(F_X^{-1}(U) \leq x) = P(U \leq F_X(x) = F_X(x) \tag{2.22}$$

It follows that

$$F_X^{-1}(U) = -\frac{\ln(1-U)}{a} \sim Exp(a) \tag{2.23}$$

In returning to step 1 from step 7, it is necessary to re-calculate only those quantities a_i, corresponding to the reactions R_i whose reactant population levels were altered in step 6; also a_0 must be re-calculated simply by adding to it the difference between each newly changed a_i value and its corresponding old value.

Since this algorithm uses M random numbers per iteration, it takes time proportional to M to update the a_i's. This results in a computational complexity that scales linearly as the problem size increases (see the example in the next section). The original Gillespie algorithm is therefore inefficient for large problems, which has prompted the development of several alternative formulations with improved scaling properties. A comprehensive review of the literature about efficient alternatives to Gillespie original algorithm can be found in [6, 90, 113, 190, 232]. The majority of the current alternatives are approximation of variables' estimates defined by the Gillespie original methods, or hybrid deterministic/stochastic algorithm (see Lecca et al. [129] for a concise but comprehensive review of stochastic and hybrid algorithms).

2.7 An Example of Real Enzymatic Reactions Simulated with Gillespie Algorithm

We give here an example of R^1 script for the Gillespie simulation of the real enzymatic activity involving fructose (F) and sorbitol (S) [22]. E denotes the active enzyme, and E_i the inactive form of E. The system involves seven reactions as follows:

$$R_1, R_2 : E + NADH \underset{k_2}{\overset{k_1}{\rightleftharpoons}} ENADH$$

$$R_3, R_4 : ENADH + F \underset{k_4}{\overset{k_3}{\rightleftharpoons}} ENAD^+ + S$$

$$R_5, R_6 : ENAD^+ + F \underset{k_6}{\overset{k_5}{\rightleftharpoons}} E + NAD^+$$

$$R_7 : E \overset{k_7}{\longrightarrow} E_i$$

where

$$k_1 = 6.2 \times 10^{-6} \text{ s}^{-1}\text{pM}^{-1}$$
$$k_2 = 33 \text{ s}^{-1}$$
$$k_3 = 2.2 \times 10^{-9} \text{ s}^{-1}\text{pM}^{-1}$$
$$k_4 = 7.9 \times 10^{-9} \text{ s}^{-1}\text{pM}^{-1}$$
$$k_5 = 227 \text{ s}^{-1}$$
$$k_6 = 6.1 \times 10^{-7} \text{ s}^{-1}\text{pM}^{-1}$$
$$k_7 = 1.9 \times 10^{-3} \text{ s}^{-1}$$

---

[1]R language software web page: www.r-project.org.

and the initial quantities of the reactants are $[E](t = 0) = 1.6 \times 10^8$ pM, $[F](t = 0) = 4 \times 10^{11}$ pM, $[NADH](t = 0) = 1.6 \times 10^8$ pM, $[S](t = 0) = [NAD^+](t = 0) = [ENADH](t = 0) = [ENAD^+](t = 0) = 0$ pM.

The library GillespieSSA [175] of R provides the functions to simulate the stochastic kinetics.

The numerical matrix of change if the number of individuals in each state (rows) caused by a single reaction of any given type (columns) is

$$
M = \begin{array}{c} \\ E \\ E_i \\ F \\ NADH \\ S \\ NAD^+ \\ ENADH \\ ENAD^+ \end{array}
\begin{array}{cccccccc}
R_1 & R_2 & R_3 & R_4 & R_5 & R_6 & R_7 \\
\left(\begin{array}{ccccccc}
-1 & 1 & 0 & 0 & 1 & -1 & -1 \\
0 & 0 & 0 & 0 & 0 & 0 & 1 \\
0 & 0 & -1 & 1 & 0 & 0 & 0 \\
-1 & 1 & 0 & 0 & 0 & 0 & 0 \\
0 & 0 & 1 & -1 & 0 & 0 & 0 \\
0 & 0 & 0 & 1 & -1 & 0 & 0 \\
1 & -1 & -1 & 1 & 0 & 0 & 0 \\
0 & 0 & 1 & -1 & -1 & 1 & 0
\end{array} \right)
\end{array}
$$

As this system is stiff as the kinetic rate constants and the abundances of the reactants range over a large interval, the simulation times are quite long. The reader can verify this by running the following R script that implements the reaction system.

```
library(GillespieSSA)

# kinetic rate constants
parms <- c(k1 = 6.2 * 10^(-6), k2 = 33, k3 = 2.2 * 10^(-9),
           k4 = 7.9 * 10^(-9), k5 = 227, k6 = 6.1 * 10^(-7),
           k7 = 1.9 * 10^(-3))

# initial state
x0 <- c(E = 1.6 * 10^8, Ei = 0, F = 4 * 10^(11),
        NADH = 1.6 * 10^8,  S = 0,  NADplus = 0,
        ENADH = 0,  ENADplus = 0)

# reaction propensities
a <- c("k1*E*NADH", "k2*ENADH", "k3*ENADH*F", "k4*ENADplus*S",
       "k5*ENADplus", "k6*E*NADplus", "k7*E")

# matrix of state-change
M <-  rbind(
            c(-1, 1,  0,  0,  1,  -1, -1),
            c(0,  0,  0,  0,  0,  0,  1),
            c(0,  0,  -1, 1,  0,  0,  0),
```

```
      c(-1,  1,   0,   0,   0,  0,  0),
      c( 0,  0,   1,  -1,   0,  0,  0),
      c( 0,  0,   0,   1,  -1,  0,  0),
      c( 1, -1,  -1,   1,   0,  0,  0 ),
      c( 0,  0,   1,  -1,  -1,  1,  0))
```

```
# run the Gillespie Direct Method
simulations <- ssa(x0=x0,a=a,nu=M,parms,tf=3000,
                   simName="Enzymatic activity", method="D")
```

```
# visualize some results
par(mfrow=c(2,1))
plot(simulations$data[,1], simulations$data[,"S"], main="Sorbit
ylab = "S (pM)", xlab="Time (s)", col="black",type="l",
lwd=2, cex.axis =1.5, cex.lab =1.5, cex.main=1.5)
plot(simulations$data[,1], simulations$data[,"NADH"], main="NADI
ylab = "NADH (pM)", xlab="Time (s)", col="black",type="l",
lwd=2, cex.axis =1.5, cex.lab =1.5, cex.main=1.5)
par(mfrow=c(1,1))
```

Running this script the user can verify that the computational times of this script are proportional to the abundance of E, and that in the case in which $[E](t = 0) \ll [NADH](t = 0)$ the stochastic simulations stops much more before than the final time "tf" (see the R code), as when $[E]$ goes to zero the reaction system dynamics stops. If $[E](t = 0) \approx [NADH](t = 0) = 10^(11)$ the running times may become long (with the initial values set in this examples the running time is about 28 min on a Desktop Windows 8.1 PC with a 3.1 GHz CPU). Furthermore if $[E](t = 0) \approx [NADH](t = 0) = 10^(11)$ but the final simulation time is long, again the simulation stops earlier than the values assigned to "tf".

The simulation time can be determined by using the function system.time in the following way:

```
system.time(
  simulations <-
ssa(x0=x0,a=a,nu=M,parms,tf=3000,
    simName="Enzymatic activity", method="D") ).
```

We leave to the reader the exercise to use this code and to make the experiments for the determination of the computational times to vary the initial quantities of reagents.

Problems

Problem 2.1. The First Reaction Method (FRM) is a variant of Gillespie Direct Method. FRM that generates a τ_k for each reaction channel R_μ ($k, \mu = 1, \ldots, M$, where M is the number of reactions in the system) according to

$$\tau_i = \frac{1}{a_i} \ln\left(\frac{1}{r_i}\right) \tag{2.24}$$

where $r_1, r_2, \ldots, r_M$ are M statistically independent samplings of $Unif(0,1)$. Then τ and μ are chosen as

$$\tau = \min\{\tau_1, \tau_2, \ldots, \tau_M\} \tag{2.25}$$

and

$$\mu = \text{the index of} \min\{\tau_1, \tau_2, \ldots, \tau_M\} \tag{2.26}$$

The steps of FRM algorithm:

1. Initialization: set the initial numbers of molecules for each chemical species; input the desired values for the M reaction constants $c_1, c_2, \ldots, c_M$. Set the simulation time variable t to zero and the duration T of the simulation.
2. Calculate and store the propensity functions a_i for all the reaction channels ($i = 1, \ldots, M$), and a_0.
3. Generate M independent random numbers from $Unif(0,10)$.
4. Generate the times τ_i, ($i = 1, 2, \ldots, M$) according to Eq. (2.24).
5. Find τ and μ according to (2.25) and (2.26), respectively.
6. Update the states of the species to reflect the execution of reaction μ. Set $t \leftarrow t + \tau$.
7. If $t < T$ then go to step 2, otherwise terminate.

Prove that the First Reaction Method is stochastically equivalent to the Direct Method. *Hint:* the random pairs (τ, μ) generated by both methods follow the same distribution.

Problem 2.2. Consider a system which can be found in various states n at time t with probability $p(n, t)$. Let $w^+(n)$ and $w^-(n)$ the rate of transition between different states. The meaning of $w^+(n)$ and $w^-(n)$ is the following: suppose to prepare N independent systems in the state n.

After a time interval dt, we will find that a proportion $a^{\pm}(n)dt$ of them has jumped into a state $n \pm 1$.

a) Write down the master equation of $p(n,t)$. Can we say that this equation is an enumerable infinite number of coupled equation (one for each n)?

b) Prove that

$$\sum_{n=-\infty}^{+\infty} w^{+}(n-1)p(n-1) = \sum_{n=-\infty}^{+\infty} w^{+}(n)p(n).$$

Hints: note that

$$\frac{d}{dt} \sum_{n} p(n,t) = \frac{d1}{dt} = 0.$$

Furthermore, change the notation from n to $n-1$, i.e. replace all n's by $n+1$.

c) Prove that

$$\frac{d\langle n \rangle}{dt} = \langle w^{+}(n) - w^{-}(n) \rangle.$$

where $\langle n \rangle$ is the mean of n. *Hints:* analogously to the previous point note that

$$\frac{d}{dt}\left[\sum_{n} np(n,t) \right] = \frac{d\langle n \rangle}{dt}$$

and

$$\sum_{n=-\infty}^{+\infty} n\, w^{+}(n-1)p(n-1) = \sum_{n=-\infty}^{+\infty} (n+1)\, w^{+}(n)p(n).$$

d) Prove that

$$\frac{d\langle n^2 \rangle}{dt} = 2\langle n(w^{+}(n) - w^{-}(n)) \rangle + \langle w^{+}(n) + w^{-}(n) \rangle.$$

e) Calculate the time derivative of the variance of n, $V = \langle n^2 \rangle - \langle n \rangle^2$.

$$\left[\text{Answer: e) } \frac{dV}{dt} = 2\langle (n-\langle n \rangle)(w^{+}(n) - w^{-}(n)) \rangle + \langle w^{+}(n) + w^{-}(n) \rangle \right]$$

Problem 2.3. Assume to have systems a reaction with a transition from state $\mathbf{x}_r$ to state $\mathbf{x}$. The vectors $\mathbf{x}_r$ and $\mathbf{x}$ have N nonnegative integer components such that $\mathbf{x}$, $\mathbf{x} \in \mathbb{Z}_+^N$. Let $\vec{n}_r = \mathbf{x}_r - \mathbf{x}$ and let the probability flow from $\mathbf{x}_r$ to $\mathbf{x}$ be quantified y the rate $w_r(\mathbf{x}_r)$. Write the master equation corresponding to R reactions satisfied by the probability p to be in state $\mathbf{x}$ at time t.

$$\left[\text{Answer: } \frac{\partial}{\partial t}p(\mathbf{x},t) = \sum_{i=1}^{R} w_r(\mathbf{x}+\mathbf{n}_r)p(\mathbf{x}+\mathbf{n}_r,t) - \sum_{i=1}^{R} w_r(\mathbf{x})p(\mathbf{x},t) \right]$$

Problem 2.4. Write a program[2] that generates the stochastic simulations for the example networks a) and b). Compare the average molecule numbers (average over, say 100 simulation runs) to the solution of the corresponding differential equations that approximate the average molecule numbers

a) An enzymatic network (with parameters in arbitrary units):

$$E + S \underset{k_2=1}{\overset{k_1=1}{\rightleftharpoons}} ES \xrightarrow{k_3=0.1} E + P$$

where the initial number of molecules is

$$[E](t = 0) = 100$$
$$[S](t = 0) = 1000$$
$$[ES](t = 0) = 0$$
$$[P](t = 0) = 0.$$

b) Gene transcription regulation network as in [151]. The network includes six species that can interact through ten reactions: the monomeric protein M that can dimerize to form the dimeric transcription factor D, the complex between DNA and D at single or double binding sites (DNA_D, DNA_2D respectively) and the messenger RNA ($mRNA$). The DNA template has two different

---

[2]The reader can use the programming language he/she prefer. However, we that Matlab (https://it.mathworks.com/products/matlab.html) and R (www.r-project.org) have libraries implementing Gillespie stochastic simulation algorithm.

binding sites for D.

$$\text{mRNA} \xrightarrow{k_1=0.043} \text{mRNA} + \text{M}$$

$$\text{M} \xrightarrow{k_2=7\times10^{-4}} \emptyset$$

$$\text{DNA\_D} \xrightarrow{k_3=0.72} \text{DNA\_D} + \text{mRNA}$$

$$\text{mRNA} \xrightarrow{k_4=3} \emptyset$$

$$\text{DNA} + \text{D} \xrightarrow{k_5=1.4} \text{DNA\_D}$$

$$\text{DNA\_D} \xrightarrow{k_6=0.48} \text{DNA} + \text{D}$$

$$\text{DNA\_D} + D \xrightarrow{k_7=1.4} \text{DNA\_2D}$$

$$\text{DNA\_2D} \xrightarrow{k_8=8\times10^{-12}} \text{DNA\_D} + \text{D}$$

$$\text{M} + \text{M} \xrightarrow{k_9=0.29} \text{D}$$

$$\text{D} \xrightarrow{k_{10}=0.5} \text{M} + \text{M}$$

where the initial number of molecules is

$$[\text{mRNA}](t = 0) = 0$$
$$[\text{M} = 2](t = 0)$$
$$[\text{DNA\_D}](t = 0) = 0$$
$$[\text{DNA\_2D}](t = 0) = 0$$
$$[\text{D}](t = 0) = 6$$
$$[\text{DNA}](t = 0) = 2.$$

Chapter 3

Graph Theory and Physics Meet Network Biology

Cellular molecules exert their functions through interactions with other molecules, which could occur in the same cell, or across cells, organs or even organisms. The complexity of the resulting interconnectivity in human is daunting with ~25,000 protein-coding genes, ~1,000 metabolites and an increasing number of newly uncovered functional RNA molecules. Network information captures knowledge on the manifold interactions (physical, regulatory, metabolic, genetic) among cellular components at the molecular level and can be an effective tool for interpreting the growing amount of biological data from genome-wide studies. The network-based approach to understand the organizing principles of cellular processes has facilitated the emergence of a body of knowledge increasingly referred to as systems biology.

This chapter provides a critical perspective on the manifold contributions of physical science into network-based approach to systems biology. In particular, the chapter (i) introduces the use of mutual information and network entropy in inferential scheme, (ii) describes the new centrality measures inspired by physics and currently used in network analysis; (iii) discusses the problem of synthetic gold-standard datasets generation for network inference performance assessment. The points (i) and (ii) are preparatory to the topics covered in the next chapter from an applicative point of view.

3.1 Physics at the Birth of Network Biology

Nowadays physicists are largely contributing to solutions of problems in biology [69]. Historically, physicists strongly contributed to the advances of molecular biology. Indeed, in the 1940s the German biophysicist, Max Delbrück described molecular genetics even before the structure of DNA had been discovered. Since then, physicists gave major contribution in structural chemistry, biochemistry, and reaction kinetics. Over the last fifteen years, physics had a significant impact on systems biology. This discipline was born around 2000, and is the result of a paradigm shift. While up to fifteen years ago, the biology focused mainly on the study of the chemical/structural properties and functions of individual genes, molecules and proteins, the biology of the past 15 years combines these studies with a more in-depth investigation on the interactions between genes, molecules, and proteins. The focus on the interactions, rather than on the structural static properties of their actors is peculiar of systems biology. Systems biology conceptualizes living systems as *networks* of interacting molecules, and makes use of the mathematical language of graph theory to describe and analyse these networks. The system-level investigation of complex phenomenology is recent in biology, but has a long tradition in theoretical physics, especially in quantum mechanics and in statistical mechanics. Furthermore, these two branches of the theoretical physics that have always had to deal with the complex interacting systems have given a significant boost to the development of computational approaches to modelling and simulation of such systems. In the same way, systems biology has led to increasingly massive use of computational approaches for understand the principles governing the dynamics of systems of interacting agents, i.e. of *dynamical networks*.

A network is an intuitive concept that systems biology uses to represent a group of interacting biological items e.g. genes, proteins, metabolites, functional complexes, enzymes. In a network-like representation of biological systems, nodes indicate the entities of the system, and pairs of nodes are connected by edges indicating the presence of a particular kind of interaction between those nodes. Network inference is the process of deduction of interactions from experimental readouts through computational analysis [94, 167]. Since networks represent blueprints of cell dynamical processes, reliable inference of large-scale causal biological interactions can contribute to a better understanding of several aspects of cell physiology, development and pathogenesis. Namely, causal topological structures of interactions within a network reflect the status (e.g. healthy or disrupted) and the collective dynamics of the components of the underlying biological process [93].

The current interest in computational methods to infer networks from quantitative experimental observations is due to the promise that these methods hold to discover novel potential relationships among system's components (e.g. regulatory relationships among genes or pathways, or biochemical interactions among proteins, metabolites, enzymes). The putative interactions inferable by a computational approach may complement and guide wet experiments which can explore just a limited sub-space of the phase space of variables. Applicative domains of biological network inference include medicine [186], drug and pharmaceutical industry [143, 235].

Beside network inference, network analysis plays an increasingly important role in the process of new knowledge discovery [4]. Network topology analysis approaches aim at identifying the organizing principles which operate in biological systems. Network centrality indexes have proven of particular effectiveness by capturing the structural attributes of a network in quantitative ways [33]. The analysis of the connectivity distribution of nodes and edges, motifs discovery and community identification have been commonly applied to biological networks for evaluation of network robustness, for pharmacology, diseases classification and, more recently, for comorbidities studies between diseases.

Nowadays, biological network inference and network analysis require a concerted expertise from multi-disciplinary sciences, e.g. physics, mathematics, computer science, biology, and chemistry, which have led to disciplines such as *computational systems biology*, whose boundaries are becoming more and more blurred. Physics has become crucial in this field because it instruments our mechanistic understanding of both static and dynamical properties of biological systems. Concepts such as mutual information and entropy [14, 31, 32, 47, 73, 153, 199], which are shared by information theory, thermodynamics, and statistical mechanics, have become part of the modern methods of network inference and analysis. Mutual information owes its popularity as measure of variable mutual dependence in inferential methods to the ability to identify non-linear relationships among nodes. Recently, mutual information-based methods showed good performances in identifying troublesome interactions in gene regulatory networks, that are known to be hard to be reverse-engineered [132, 218, 226, 236]. Beyond popular centrality measures, such as degree, betweenness and clustering coefficient, assortativity and closeness, new indexes such as vibrational centrality and resistance distance [72, 74, 77] have been lately inspired by thermodynamics and statistical mechanics concepts in order to assess network vulnerability, i.e. network inertia to a stress. In fact, biological systems are highly dynamic entities that must continuously respond to a host of environmental and genetic changes, which are independent of the organizational architecture of the network. In these cases, standard centrality indexes, which entirely rely on network structural information, do not convincingly show to be able to express network behaviour in the presence of dynamical conditions. From this standpoint, applying vibrational centrality and resistance distance is proving more suitable for learning the property of network vulnerability. Namely, centrality indexes inspired to thermodynamics and statistical mechanics quantify the vulnerability of nodes to external stresses, i.e. by stresses that are independent of the organizational architecture of the network. Changes in the physiological conditions of the cell, such as pH and/or concentration of chemicals are examples of this kind of stresses. Vibrational centrality defines the vulnerability of nodes as function of their degree and of the strength of their edges [180], and resistance distance defines a graphical distance of pairs of nodes. Accounting for both the number of connections and for the intensity of the connections, they may better reflect the robustness of nodes to multiple edge failures or nodes removal and/or knock-down.

In the vibrational centrality approach [77] the external stresses a system may be exposed to are accounted for by the concept of temperature. Herein temperature is meant to be a metaphor of all the different types of stress that the network can be submitted to. In line with this metaphor, nodes are rigid spheres and edges are elastic springs, submerged in a thermal bath at a given temperature [77, 180]. Vibrational centrality quantifies the magnitude of the "oscillation" of a node in response to a stress, and thus it estimates the tendency of the node to resist a change in motion, namely its robustness.

As to the resistance distance [76], networks are pictured as electrical circuits where edges are replaced by resistors and nodes are connected to batteries [55]. Resistance distance is also a measure of network vulnerability, as it is closely linked to the speed of propagation of perturbations from node to another. The effects of a stress on a node propagate faster to closer nodes that to distant nodes. Namely, a map of the resistance distances between couples of nodes could individuate network regions of fast and low paths of propagation of perturbation. Paths along which stresses spread quickly are expected to be the most vulnerable parts of the network. Vibrational centrality provides information about the propagation of stimuli from a node to other nodes of the network. In fact, vibrational centrality can be shown to be consisting of two terms: the amount of stress adsorbed by the node and the amount of stress reflected to the node and propagated to the other.

Although the contribution of physical models to computational biology has undoubtedly been beneficial, a critical approach and a conscious application of physical models to network biology should always be adopted. To bring examples of the necessity of these precautions we will discuss the issue of network inference validation. As soon as a critical mass of new network inference methodologies were published and tools were available to the community, the problem of a comparative assessment of their performances became of great importance. Current performance assessment methods involve the application of the algorithm to benchmark datasets and the comparison of the network predictions against the gold standard or reference networks. The generation of gold-standard datasets is made by simulating the dynamics of the network via differential equations inspired by the law of mass actions (i.e. Newton laws). The solution of systems of this kind of equations generates datasets of indirectly correlated variables (nodes) that the network inference methods reverse-engineer as a partial (if not total) transitively closed graphs. In this chapter we will provide examples of how the use of rate equations derived from the Newtonian dynamics, could provide unsatisfactory gold-standards.

3.2 Mutual Information-Based Network Inference

The rationale of mutual information-based network inference is to infer a link between a couple of nodes if it has a high score based on mutual information. Mutual information-based network inference proceeds in two steps. The first step is the computation of the mutual information matrix (MIM), a square matrix whose i, j-th element

$$M_{ij} = I(X_i; X_j) = I(X_j, X_i) =\equiv \int_{X_j} \int_{X_i} p(X_i, X_j) \log \frac{p(X_i, X_j)}{P(X_i)p(X_j)} \quad (3.1)$$

is the mutual information between the variable X_i and the variable X_j. $p(X_i, X_j)$ is the joint probability density function of X_i and X_j, and $p(X_i)$ and $p(X_j)$ are the marginal probability density functions of X_i and X_j respectively. If the log base 2 is used, the unit of mutual information is the bit. Mutual information measures the information shared by X_i and X_j by estimating how much knowing one of these variables reduces uncertainty about the other. For example, if X_i and X_j are independent, then knowing X_i does not allow to know X_j and vice versa, so their mutual information is zero. At the other extreme, if X_i is a deterministic function of X_j, knowing X_j determines the value of X_i, and as a consequence the mutual information is the uncertainty contained in X_j alone, namely the *entropy* of X_j. In this context, the term *entropy* usually refers to the Shannon entropy $H(X_j)$, which quantifies the expected value of the information "contained" in a variable:

$$H(X_j) = - \int_{-\infty}^{\infty} p(X_j) \log P(X_j) dX_j$$

The dependence between X_i and X_j is quantitatively proportional to the ratio between the joint probability of X_i and X_j, $P(X_i, X_j)$, and the product of the single probabilities $p(X_i)$ and $p(X_j)$. If X_i and X_j are independent, then the joint probability is equal to the product of the single probabilities so that the argument of the logarithm is equal to one and the logarithm is zero.

Mutual information provides a generalization of the correlation since it is a non-linear measure of dependency. A distinctive advantage of using MIM is the ability to deal with up to several thousands of variables, also in the presence of a limited number of samples, because MIM computation requires only estimations of a bivariate mutual information term, i.e. the mutual information between couples of variables.

The second step is the computation of an edge score for each pair of nodes by an inference algorithm that takes the MIM matrix as input [11]. ARACNE [149] is a well-established network inference tool and paved the way to several other methods based on mutual information, and mostly tailored to the inference of gene regulatory networks, rather than of metabolic and signalling networks. The most recent methods are MINET [153], the work of K-C. Liang et al. [132], V. Chaitankar et al. [47], X. Zhang et al. [236] that introduced the three-way mutual information to detect the joint regulations of a target gene by two or more genes.

Since mutual information is a symmetric measure, it is not possible to derive the direction of an edge using a mutual information network inference technique. Nevertheless, the orientation of the edges can be obtained by using procedures like inductive causation algorithms [36]. The performances of mutual information based networks are satisfactory on real data and compete with the current other methods on synthetic data, as we will discuss in more detail in Section 3.5.

3.3 Thermodynamics Applications in Biological Network Analysis

Thermodynamics has borrowed a number of physical concepts to the study of network properties in response to external stimuli (e.g. environmental and genetic changes, pharmacological treatments). Here we discuss two concepts which are firmly ingrained in complex network analysis: vibrational centrality, network entropy.

3.3.1 *Vibrational Centrality*

Complex networks are continuously exposed to stresses which are independent of the organizational architecture of the network. Examples of stresses are changes in the physiological conditions in a cell, such as pH, physical temperature or chemicals concentration. To measure the susceptibility of network nodes to external stresses, a novel centrality measure, termed *vibrational centrality*, was recently introduced by E. Estrada [77]. The definition of this index is based on the analogy of a network with a physical system in which the nodes are rigid spheres and the edges are springs (Figure 3.1). The effects of stresses on such a system are modelled as nodes displacements, i.e. nodes reactions to changed conditions could be represented in terms of nodes vibrations, that are deviations of nodes positions from their equilibrium ones.

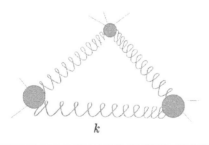

k

Figure 3.1: A physical analogy of a graph with a physical system of nodes linked by springs. Nodes of networks are abstracted by harmonic oscillators [72].

In order to define the position of nodes, the network is embedded into an n-dimensional Euclidean space (n being the number of nodes in the network) represented by the Moore-Penrose pseudo-inverse of graph Laplacian $L = D - A$, where D is the diagonal matrix of degrees and A is the graph adjacency matrix. Henceforth we denote by L^+ the pseudo-inverse of L. Each diagonal entry of L^+, denoted as l_{ii}^+ for the i-th node, represents the squared distance of node i to the origin in this n-dimensional space and provides a measure of the nodes topological centrality, given as

$$C(i) = \frac{1}{l_{ii}^+}. \tag{3.2}$$

Closer the node i is to the origin in this space, or equivalently lower the l_{ii}^+, more topologically central node i is [180].

A network subjected to some sort of external stress can be modelled by a network submerged in a thermal bath at temperature T. This abstraction allows to define the vibrational potential energy of the network as it is usually defined for a system of microscopic particles, i.e.

$$V(x) = \frac{k}{2}x^T L x \tag{3.3}$$

where k is the spring constant, and x is the vector whose i-th entry is the displacement x_i of the i-th node. x^T is the trasponse of vector x. According to the metaphor of a system submerged into a thermal bath, the probability distribution of the displacement of the nodes is given by the Boltzmann distribution

$$P(x) = \frac{e^{-\frac{1}{T}V(x)}}{Z} = \frac{1}{Z}\exp\left(-\frac{k}{2T}x^T L x\right) \tag{3.4}$$

where the partition function Z of the network is defined as follows

$$Z \equiv \int dx \exp\left(-\frac{k}{2T}x^T L x\right).$$

Given $P(x)$, the mean displacement of a node i, is by definition

$$\langle \Delta x_i \rangle \equiv \sqrt{\int x_i^2 P(x) dx} \tag{3.5}$$

It can be shown [74]

$$\langle \Delta x_i \rangle = \sqrt{\frac{T}{k}(L)_{ii}} \tag{3.6}$$

Now, lets go back to the physical world. If the stresses are too high, i.e., high temperature (i.e. $T \to \infty$), then all the interactions in the network are interrupted, the network is destroyed, which is a very realistic picture of what happens if the physiological conditions are extreme. Now, consider a biological network submitted to the normal fluctuations in which a cell lives. In this case, there are some nodes that "transmit" more than others the perturbations that they feel from the environment. That is, if the physiological conditions of a cellular network change, then all the nodes "feel" those changes, but some are more affected than others. This is reflected by the vibrations of the nodes, which represent the deviations of the positions of a node from its equilibrium position. Let us consider that the equilibrium position for a given protein is just its state under normal physiological conditions. Now, if these conditions are altered this protein changes from that state-state position to adapt to the new conditions. This change is accounted for the vibrational centrality.

Equations (3.3) and (3.4) describe the potential vibrational energy of a network submerged into a thermal bath at the temperature T. In these equations, the thermal bath represents here an external change which affects all the links in the network at the same time. It fulfils the requirements of being independent of the topology of the network and of having a direct influence over it. Then, after equilibration all links in the network will be weighted by the parameter $\beta \equiv 1/T$. When the temperature tends to zero, all edges have infinite weight. Such a graph has resemblances with the solid state of a matter, since a solid is formed by particles that are almost at rest and strongly linked one to another. In the other extreme, when the temperature tends to infinity, all edges have zero weights, i.e. the graph has no edges. Such a graph is commonly referred to as *empty* graph. An empty graph is similar to a gas submitted to very high temperature. Under this condition the gas is formed by free particles that do not interact with one another. In between these two configurations, there is the case in which $T = 1$. This case corresponds to the *strict* graph in which every pair of connected nodes has a single link, i.e. the graph has no loops and nodes do not have multiple edges. The strict graph may be then analogous to the liquid state of the matter.

For the case of trees, it is easy to prove that vibrational centrality can be expressed by two terms as in Eq. (3.7) [72].

$$(\Delta x_i)^2 = \frac{2n-1}{2n^2}\left[s_i - \left(\frac{1}{2n-1}\right)\sum_{q \neq i} s_q\right]. \tag{3.7}$$

where s_i represents the sum of all the distances that information has to travel in order to reach all nodes staring from i, while the term $-\sum_{q \neq i} s_q$ estimates the information which is reflected from all nodes that have received it from node i.

Finally, in Figure 3.2 we show the behaviour of vibration centrality versus the standard centrality measures. Vibrational centrality decreases as degree, betweenness, closeness, subgraph centrality, eigenvector centrality and information centrality increase. Vibrational centrality is not instead correlated to clustering coefficient. The plots have been obtained for a scale-free random graph of 2,000 nodes and 35,000 arcs.

Vibrational centrality (VC)

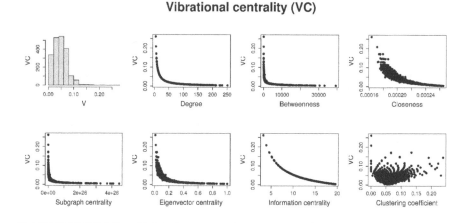

Figure 3.2: **Vibrational centrality against canonical topological centralities measures. The plots have been obtained for a scale-free random graph of 2,000 nodes and 35,000 arcs. The degree distribution of this network follows a power-law distribution with a power-law coefficient equal to 2.2.**

Vibrational centrality was found to achieve high resolution in the identification of the most vulnerable nodes in a network, compared to standard topological measures such as node degree, which can account for only the nearest-neighbours of a node to quantify the relative contribution of the node to network integrity [77].

3.3.2 Network Entropy: A graph theory based definition

The robustness of a dynamical system has been also characterized in terms of network entropy. In this case, network robustness is derived from the uncertainty in the variables associated with the network constituents.

The spectrum of a network, i.e. the density of the eigenvalues of its adjacency matrix, provides information on the possible states of a network. Indeed, the eigenvector associated with each eigenvalue of the adjacency matrix represents a state where the network exists and the pattern of positive/negative signs of the nodes along the eigenvector describes the partition of nodes in this state. For instance, Figure 3.3

shows the network spectrum and nodes density for the first two eigenvectors of a random graph with two disconnected communities.

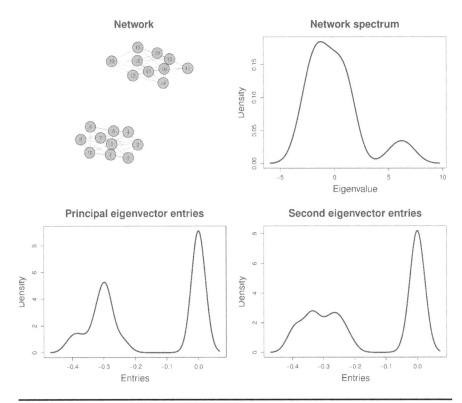

Figure 3.3: The density of the eigenvalues of the adjacency matrix, known as network spectrum, informs of the presence of communities in the network. The figure shows the spectral density and the entry densities for the eigenvectors corresponding to the first and second largest eigenvalues for a random graph with two disconnected dense groups of nodes. Each eigenvector of the network's adjacency matrix corresponds to a possible state of the network. Partitions of nodes in communities are determined by the sign of the entries of the eigenvectors of the adjacency matrix [72]. This example shows nodes yielding positive and negative entries for the eigenvector corresponding to the second largest eigenvalue. Therefore, this eigenvector specifies a bipartition of the network. These plots have been obtained for a random graph of 20 nodes generated according to the Erdös-Rényi model.

The j-*th* state of the network corresponds to the eigenvector v_j associated with the eigenvalue λ_j of the adjacency matrix of the network. There are many possible physical interpretations for the states of a network. For example, the sign patterns in the entries of the eigenvectors correspond to different vibrational modes of the nodes. A simple example of vibrational mode in this mechanical analogy is the mode defined by the zero eigenvalue of the Laplacian matrix. The zero eigenvalue corresponds to the mode where all the nodes move coherently in the same direction, so that the whole network moves in one direction. Namely, the zero eigenvalue corresponds to the motion of the centre of mass, so that its does not contribute to the vibrational energy of the network.

The analogy with statistical mechanics goes further and introduces the definition of a Schrödinger operator of a network when the potential energy operator is given by the node degree. Physical interpretations for the states of a network can be better understood by introducing the definition of a Schrödinger operator of when the potential energy operator is given by the node degree.

$$V_i = -d_i. \tag{3.8}$$

In this case, the Hamiltonian of the network is the negative of the adjacency matrix

$$H = -A \tag{3.9}$$

and the energy of the j-*th* state of the network is

$$E_j = \lambda_j. \tag{3.10}$$

Therefore, the state representing an unipartition of the network has a null energy and thus is the most stable state. A network bipartition corresponds to the second most stable state of the network, and so forth. The probability that the system is found in the j-*th* state is [28, 72, 73]

$$P_j = \frac{e^{-\beta E_j}}{\sum_j e^{-\beta E_j}} = \frac{e^{-\beta \lambda_j}}{\sum_j e^{-\beta \lambda_j}} \tag{3.11}$$

where Z is again the partition function of the network. The quantity at the denominator is the generalized Estrada index [72]

$$EE \equiv \sum_j e^{-\beta \lambda_j} = \mathrm{Tr}(\exp \beta A) \tag{3.12}$$

The network information theoretic entropy S is now defined by the Shannon expression as follows:

$$S = -k_B \sum_j \left[P_j \left(\beta \lambda_j - \log EE \right) \right]$$

$$= -k_B \beta \sum_j \lambda_j P_j + k_B (\log EE) \sum_j P_j \tag{3.13}$$

The Shannon entropy E of the weights of edges incident to a node is also know as *node diversity*. Let d_i the total degree of vertex i, and w_{ij} the weight of the edge(s) between vertices i and j, then the node diversity is defined as follows.

$$E_i = \frac{H_i}{\log(d_i)} \tag{3.14}$$

where

$$H_i = -\sum_j (p_{ij} \log(p_{ij})), \quad j = 1, \ldots, d_i$$

$$p_{ij} = \frac{w_{ij}}{\sum_l w_{il}}, \quad l = 1, \ldots, d_i$$

In Figure 3.4, we report the behaviour of graph diversity against the standard centrality indexes. The plots show absence of correlation between graph diversity and the other centrality measures. Figure 3.5 shows the distribution of graph diversity for scale-free random graphs and non scale-free random graph (obtained with the Erdös-Rényi algorithm). All the graphs have 1,500 nodes. The scale-free one has 5,000 edges, and the number of edges of Eröds-Rényi graphs are determined by the probability p for drawing an edge between two arbitrary vertices. In this simulation we explored the range of probability from 0.1 to 1 by steps of 0.2. For non scale-free random dense graphs, having an

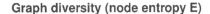

Graph diversity (node entropy E)

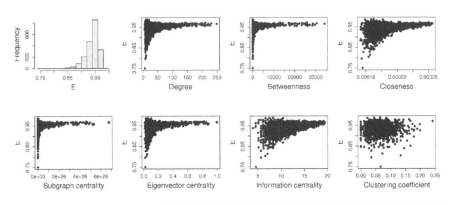

Figure 3.4: No correlation exists between graph diversity and the standard centrality measures. The plots have been obtained for a scale-free random graph having 2,000 nodes and 25,000 arcs. The power-law exponent has been set to 2.2.

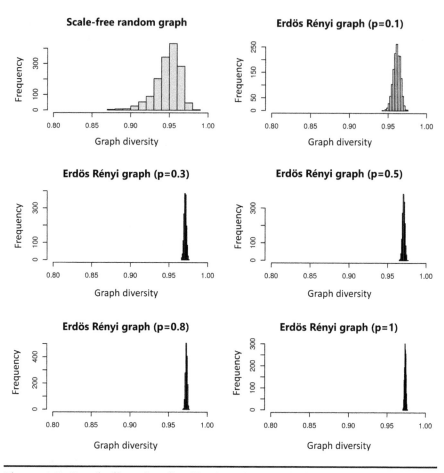

Figure 3.5: Graph diversity distributions in a scale-free random graph and in non scale-free Erdös-Rényi random graphs at different values of the probability p for drawing an edge between two arbitrary vertices.

almost uniform distribution of degree, the range of variability of entropy is narrower than that in scale-free graphs. Furthermore, note that the distribution of graph diversity for non scale-free graph narrows as p increases.

3.4 Electronic Physics Applications in Network Analysis

The conventional distance between two nodes, known as *weighted shortest path length*, is taken as the minimal sum of (positive) edge weights along the path connecting the two nodes. A more general distance measure has been introduced by Klein and Randič [125] to account for all the paths connecting two nodes. This new network metric is formalized through the concepts borrowed by electrical network theory, wherein a fixed resistor is imagined on each edge. Then, the distance between two nodes is defined as the (effective) resistance between the two nodes when a battery is connected across them. Thus, to identify the distance between nodes A and B, we could imagine the corresponding electrical circuits shown in Figure 3.6.

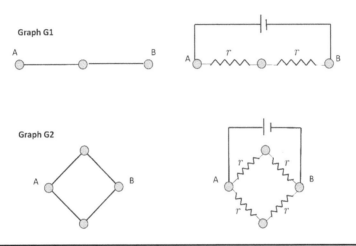

Figure 3.6: Examples of graphs represented as electrical circuits. A resistor is introduced on each edge and a battery is linked between A, B-pairs of nodes.

If all the resistors take a value of r Ohm, according to the Kirchhoff's laws, the effective resistance between A and B is:

$$\text{G1:} \quad \Omega_{AB} = r + r = 2r$$

$$\text{G2:} \quad \Omega_{AB} = \left[\frac{1}{r+r} + \frac{1}{r+r} \right]^{-1} = r$$

From this we see that the number of paths as well as their lengths contribute to the effective resistance. In particular, the reisistance distance between two nodes decreases as the number of routes from a node to the other increases. Having k parallel edges/paths leads to an effective resistance that is decreased by $1/k$.

The effective resistance between other pairs (i, j) of nodes are obtained by de-taching the battery from A and B and re-attaching it between i and j, and using again the Kirchhoff's rules. A characteristic of effective resistance Ω_{ij} is that it is a non-decreasing function of the edge resistance (this result refers to as "Rayleigh's Monotonicity Law"). Furthermore Klein and Randič [125] proved that the effective resistance is a distance, that they termed *resistance distance*, and that the conventional and resistance distance are the same between every pair of nodes of a connected graph if and only if the graph is a tree.

Resistance distance can be expressed in term of network Laplacian L as follows

$$\Omega_{ij} = \Gamma_{ii}^{-1} + \Gamma_{jj}^{-1} + 2\Gamma_{ij}^{-1} \tag{3.15}$$

where $\Gamma = L + I \cdot \frac{1}{n}$, and n is the number of nodes of the network. In Figures 3.7 and 3.8, we see that resistance distance does not correlate either to unweighted shortest path length or weighted shortest path length.

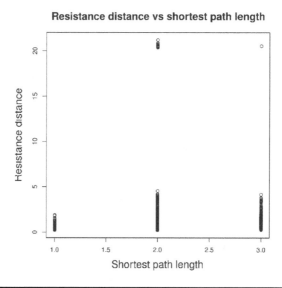

Resistance distance vs shortest path length

Figure 3.7: Resistance distance versus unweighted shortest path length. The two measures are not correlated. Different values of resistance distance are associated to shortest path having the same unweighted length.

Finally, it can be seen that resistance distance is related to the vibrational centrality as follows [72]:

$$R_i = n(\Delta x_i)^2 + \frac{K_f}{n} \tag{3.16}$$

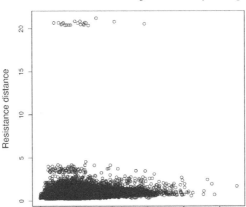

Figure 3.8: Resistance distance versus weighted shortest path length. The two measures are not correlated.

where R_i is the sum of resistance distances for the node i and K_f is the Kirchhoff index of the network given by $\sum_{i=1}^{n} \sum_{j=1}^{n} \Omega_{ij}$, where Ω_{ij} is the resistance distance matrix of the network.

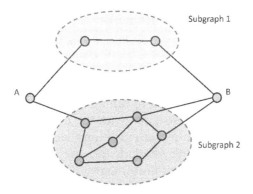

Figure 3.9: Selection of the subgraph relevant to give the most detailed information about the interaction between nodes A and B can be performed by computing the resistance distance of the subgraphs. For instance, if the resistance values on each edge is greater than one, Subgraph 2 is selected as the most relevant, as it is the one with the minimal resistance distance, i.e. with the largest number of routes between A and B.

Resistance distance has recently found many applications in bio-chemistry [215], where it used in algorithms of subgraph extraction. A typical problem in biochemical networks is to identify a relevant subgraph of the network, which best represents the relations between k given nodes of interest of the network. S. Vast et al. [215] report a practical instance of this problem: assume we would like to analyse the synthesis of pyruvate from glucose and would like to study the possible influence of the expression of a given gene on phospho-fructokinase-2, in the context of the regulation of this metabolic pathway. In this case, we have k nodes of interest in a possibly very large graph of interactions and we would like to extract a relevant subgraph explaining the relations between these k nodes.

Resistance distance can be used as a goodness function to select relevant subgraphs. Consider for example, the graph in Figure 3.9, and $k = 1$ nodes (A and B). A relevant subgraph of the connections between A and B could be selected on the criterion of minimal resistance distance.

3.5 Assessment of Network Inference Methods and the Issue of Generation of Gold-Standard Data

Since nowadays several network inference methods are available, an accurate and systematic assessment of their performances is needed. However, the performance evaluation is hampered by the difficulty of constructing adequate benchmarks and the lack of tools for a comparative analysis of network predictions on such benchmarks [192]. In this review we focus on the first of these two aspects. The process of benchmark generation consists of two steps: firstly, a topology is established, and, secondly, quantitative synthetic data (time-series or steady state data) are generated to mimic the experimental measures of the quantitative features associated with the nodes (e.g. gene expression level, protein or metabolite concentrations, and so forth). In order to produce synthetic data, the network topology must be endowed with dynamical models. Systems of non-linear ordinary differential equations (ODEs) are widely used [96, 185] for this purposes. GeneNetWeaver (GNW) is currently the most accredited tool for *in silico* benchmark generation and performance profiling of network inference methods [192].

GNW is specialized in the generation of synthetic data for gene regulatory networks. In GNW known transcriptional networks (e.g. *Escherichia coli*, *Saccharomyces cerevisiae*) are translated into detailed dynamical models of gene regulation accounting for both transcription and translation, as well as molecular and measurement noise. Both transcription and translation are modelled using a standard thermodynamic approach in which for each gene i of a network, the rate of change of mRNA concentration F_i^{rna} and the rate of change of protein concentration F_i^{prot} are described by

$$\frac{dF_i^{rna}}{dt} \equiv \frac{dx_i}{dt} = m_i f_i(y) - \lambda_i^{rna} F_i^{rna} \tag{3.17}$$

$$\frac{dF_i^{prot}}{dt} \equiv \frac{dy_i}{dt} = r_i x_i - \lambda_i^{prot} F_i^{prot} \tag{3.18}$$

where m_i is the maximum transcription rate, r_i the translation rate, λ_i^{rna} and λ_i^{prot} are the mRNA and protein degradation rates. $f_i(\cdot)$ is the input function of gene i [148]. The input function computes the relative activation of a gene, which is between 0 (when the gene is shut off) and 1 (when the gene is maximally activated), given the concentrations of the transcript factors $y = \{y_1, y_2, \dots\}$. The input function is expressed in terms of probability that a gene is a state S. For simplicity suppose that a gene i is regulated by a single transcription factor j. In this case its promoter has two states: either the transcription factor is bound (state S_1) or the transcription factor is not bound (state S_0). The probability that a gene is in state S_1 at an instant in time is given by the fractional saturation, i.e.

$$P\{S_1\} = \frac{\nu_i}{1 + \nu_1} \quad \text{with} \quad \nu_i = \left(\frac{y_j}{k_{ij}}\right)^{n_{ij}}$$

where k_{ij} is the dissociation constant and n_{ij} is the Hill coefficient. Given $P\{S_1\}$ and is complement $P\{S_0\}$, the input function is

$$f(y_i) = a_0 P\{S_0\} + a_1 P\{S_1\}$$

where a_0 and a_1 are the relative activations, when the transcription factor is unbound or bound respectively. If a gene is controlled by N transcript factors, then it can be in 2^N states, as each of the transcription factor can be bound or not bound. The input function $f(y) = f(y_1, y_2, \dots, y_N)$ is given by

$$f(y_1, y_2, \dots, y_N) = \sum_{m=0}^{2^N - 1} a_m P\{S_m\}.$$

In the case a gene has two inputs, Marbach et al. [148] report that the input function is

$$f(y_1, y_2) = \frac{a_0 + a_1 \nu_1 + a_2 \nu_2 + a_3 \rho \nu_1 \nu_2}{1 + \nu_1 + \nu_2 + \rho \nu_1 \nu_2}$$

with $\nu_j = (y_j/k_j)^{n_j}$, where k_j is the dissociation constant, ρ the cooperativity factor, and a are the relative activations.

If two genes x_1 and x_2 have the same input y_1, then dx_1/dt and dx_2/dt are correlated, because

$$\frac{dx_1}{dt} = m_1 f_1(y_1) - \lambda_1^{RNA} x_1$$

$$\frac{dx_2}{dt} = m_2 f_2(y_1) - \lambda_2^{RNA} x_2$$

with

$$f_1(y_1) = f_2(y_1) = a_0 P\{S_0\} + a_1 P\{S_1\} = \frac{a_0 + a_1 \frac{y_1}{k_1}}{1 + \frac{y_1}{k_1}}.$$

The prediction of a link between x_1 and x_2, if occurs, is a false positive; an edge, that has not been designed in the topology, is scored by a non null correlation. Therefore, the appearance of these spurious edges will invariably reflect on the benchmarking of any network inference methodology.

By solving systems of ODEs and simulating also (different) models of experimental noise, GNW computes a detailed simulation of the mRNA and protein concentration level of each gene. GNW is also the tool used by the DREAM initiative (Dialogue on Reverse Engineering Methods) [202] to provide the synthetic data for the network inference annual challenge.

ODE-based approaches to the generation of benchmark datasets are used also for specifying dynamic model of metabolic networks where rate equations are used to model chemical interactions among molecules. Examples of such systems of equations are shown in Table 3.1. However, in spite of fact that the ODE-based specifications are well performing models of the dynamics, and have well-established foundations in Newton mechanics, that make them suitable for the description of the kinetic theory of molecular collisions in chemical reactions, these models must be used with caution for the generation of synthetic data for inference assessment. In support to this recommendation, we report here three simple examples. We figured out three network topologies, Network 1, Network 2, and Network 3 as in Figure 3.10, and an elementary Michaelis-Menten kinetic as in Figure 3.12. We simulated the corresponding dynamics models, reported in Table 3.1, to obtain the data from which to reverse-engineer the topologies. The pairwise calculation of Pearson correlation, mutual information (Figures 3.10 and 3.11) and conditional entropy shows that a number of indirect effects, not designed in the topology, are scored by high values of correlation, mutual information and conditional entropy. Indirect effect not explicitly conceived in the topology are determined by the equations of the dynamic model. For instance, sister nodes share the same functional form of the dynamics. In particular, in the simplest case, the functional form of the dynamics of a sister node s is directly proportional to the dynamics of the other sister node. Consider a simple network of three nodes, A, B, and C, such that $A \to B$ and $A \to C$. Applying the law of mass action we have that $B(t) = k_1 \int A(t)dt$, $C(t) = k_2 \int A(t)dt$, and thus $B(t) = (k_1/k_2)C(t)$.

Table 3.1: Law of mass action rate equation used for the generation of the synthetic gold-standard data of Network 1, 2, a,d 3 showed in Figure 3.10. In networks 1, the initial amounts are set as follows: $A = 100$, $B = C = D = E = F = 0$; in network 2, $A = B = C = D = E = F = 1$; in network 3, $A = B = D = 1$. In Michaelis-Menten equations, they are set as follows: $A = 1$, $B = 20$, $C = 0$, and $D = 0$.

Rate	Network 1	Network 2	Network 3
$\frac{dA}{dt}$	$-k_1 A$	$-k_1 A$	$k_1 B - k_{-1} A$
$\frac{dB}{dt}$	$k_1 A - (k_2 + k_3)B$	$-k_2 B$	$k_2 D$
$\frac{dC}{dt}$	$k_2 B$	$k_2 B$	0
$\frac{dD}{dt}$	$k_3 B - (k_4 + k_5)D$	$-k_4 D$	0
$\frac{dE}{dt}$	$k_4 D$	$k_4 D$	0
$\frac{dF}{dt}$	$k_5 D$	0	0
	$k_1 = 0.01, k_2 = 0.2, k_3 = 0.9, k_4 = 0.000\ k_5 = 0.03$	$k_1 = 0.1, k_2 = 1, k_4 = 1$	$k_1 = k_{-1} = k_2 = 1$

Michaelis-Menten network			
$\frac{dA}{dt}$	$-k_f * A * B + k_r * C + k_{cat} * C$		
$\frac{dB}{dt}$	$-k_f * A * B + k_r * C$		
$\frac{dC}{dt}$	$k_f * A * B - k_r * C - k_{cat} * C$		
$\frac{dD}{dt}$	$k_{cat} * C$		
	$k_f = 0.01, k_r = 0.0005, k_{cat} = 0.002$		

Typically, the disruption of undesired indirect effects in gold-standard generation via ODEs is achieved by adding noise to the data we want to be uncorrelated, but, this solution is not always optimal or viable, since it is not easy to keep under control the propagation of this uncorrelation to links that we designed in the topology and we want to be kept by the inference algorithm.

As a consequence of all of this, mutual information-based methods, which usually performing very well on real data, might show a decrement of their performances on data synthetically generated via ODEs model. Often, preliminary performance tests of network inference algorithms on synthetic data push the authors to equip their method with a procedure for the elimination of effects generated by the models. Although the spurious edges are determined by the equations, they are considered as *false positives*, since they are not explicitly present in the designed gold-standard topology.

Here we report recent studies that analyse network inference method performance on synthetic and real data (both dynamic and static data), showing lower accuracy and sensitivity on synthetic dataset, even for mutual information based approaches. Lopes et al. [137] conducted an experimental assessment of static and dynamic algorithms for gene regulation inference from time series gene expression microarray and did not rely on synthetic data. 500 networks of 5 genes were inferred for three species (*E. coli*, yeast and fruit fly). Gold standard was defined as being interactions documented in the literature. An area under the precision recall curve (AUPRC) was assigned to each inferred network. The AUPRC values of the 500 networks predicted by an inference method were averaged, and this value was used to score the method. Among the tested dynamic models, some methods were based on mutual information (TD-ARACNE [237], time lagged MRNET [152] and time lagged Context Likelihood of Relatedness [48, 80]) while others (Simone, GI DBN) were not. Relative to each validation dataset, methods relying on mutual information yield AUPRC values comparable or significantly better than dynamic models relying on other features.

P. Zoppoli et al. [237] provide a comparison of performances of TD-ARACNE (TimeDelay-ARACNE) on simulated gene expression data according to stochastic difference equations as well as on *in vivo* yeast synthetic network (IRMA). The authors compared TD-ARACNE with BANJO [97] and TSNI [19] based on Bayesian networks and ordinary differential equations, respectively. Beyond the general better score achieved by TD-ARACNE in almost each validation setting, it is to note that the performances of each reverse-engineering algorithm were superior by using the *in vivo* synthetic network than by using simulated data. This observation strengthens the usefulness of in vivo synthetic networks for assessing reverse engineering.

Finally, A.F. Villaverde et al. [218] compare MIDER, a tool combining mutual information and entropy reduction, with other reverse engineering approaches based on information theoretic concepts by using seven benchmark setting, two of which consisted of DREAM 4 [201] *in silico* gene networks. The authors showed that the performances of all the methods on the *in silico* gene networks were poorer compared with the other benchmarks.

All this shows that first, to generate optimal gold-standard to benchmark network inference predictions is still an open question, and, second, network inference algorithms have to deal carefully with criteria and methods to discover indirect interactions and classify them as potential false positives. In fact, sophisticated network inference algorithms typically implement different approaches to try to distinguish direct from indirect interactions and co-regulation, like, for instance, the recent method of network deconvolution of Feizi et al. [82] and the global silencing method of Barzel et al. [23]. These approaches use different criteria to detect indirect effect, but at the end, all of them eliminate them from the network. While on synthetic topologies and data it is easy to assess the performance of these methods for the reduction of indirect interaction, on real data, it could be more complicated. In real *in vivo data*, not all indirect effects can be classified as false positives and scored to be eliminated. They could be real observed interactions (i.e. highly scored true positives). The lesson we learn from this is that network inference methods even when show good performances on synthetic data, have to be considered more as generators of hypothesis on the wiring diagram of interactions underlying a biological process, rather than providers of a single accurate model.

3.6 Network Biology is Transformed by Physics

Current biology is transformed by physicists and chemical physics, who apply powerful concepts to infer and analyse mechanisms of biological processes. In particular, systems biology benefits of the convergence between biology and physics. In this review, we focused on the contributions of physics to network biology, that is the main area of system biology. Indeed, a network-like (or graph-like) representation of a biological system suggests the application of physical and information theory concepts to infer its topological structures from experimental data, and to analyse the topological properties of this structures. Concepts developed by statistical mechanics (who rely on information theory), thermodynamics, and electrical physics hold the promise to be simple yet powerful tools to deduce models of biological process and to analyse the properties both of the actors of these process (e.g. genes, proteins, metabolites, etc) and the properties of their interactions (e.g topological distribution and intensity of the interactions). The hot question that network analysis is currently trying to address is related to the determination of the vulnerability of the nodes network. The

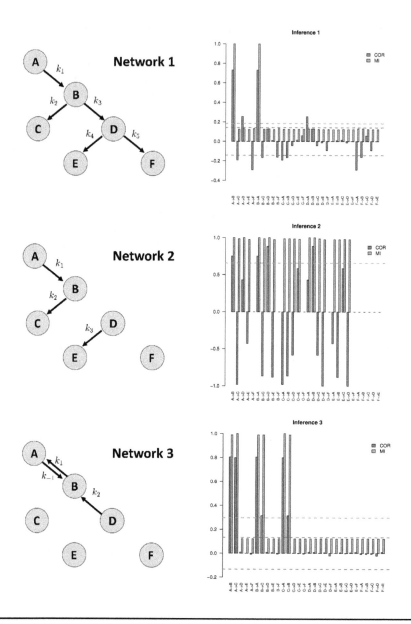

Figure 3.10: Correlation (COR) and mutual information (MI) between nodes of three simple network models (Network 1, Network 2, and Network 3). Synthetic data for the inference have been generated by the simulation of the dynamical models reported in Table 3.1. Both initial amounts of variables and rate constants are expressed here in arbitrary units.

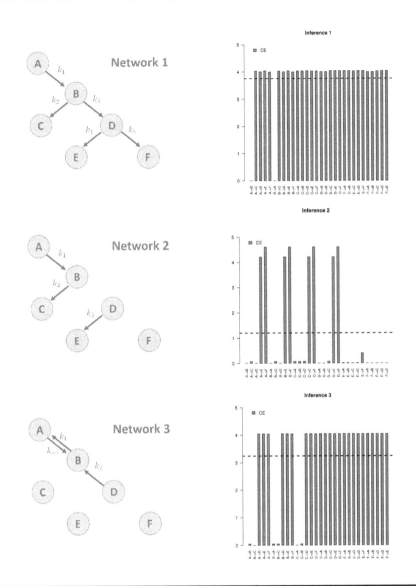

Figure 3.11: Conditional entropy (CE) between nodes.

importance of this question stay in the potential impacts that its answers can have in the design of drugs and therapeutic schemes (e.g. dosage and timing). In this ambit physics proposes models of particle interacting systems borrowed by thermodynamics and electrical physics. Vibrational centrality and resistance distance are the newest example of centrality measures expressing the strength of the response of a network to perturbations.

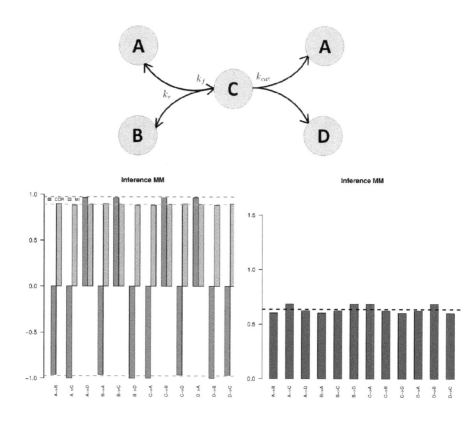

Figure 3.12: Correlation, mutual information and conditional entropy between nodes of a Michaelis-Menten (MM) reaction network.

Finally, we close up the chapter by discussing the application of physical chemistry models to the assessment of the performance of computational methods of network inference. We show that the widely used techniques to generate synthetic gold-standard data and networks based on ordinary differential equations inspired to physical law of mass action do not give optimal benchmarks. We analysed this problem with the support simple examples and tried to suggest possible solution toward the generation of better gold-standard generation. This last argument of our review likes to advertise the reader also of the limitations that in some contexts the application of physical models to biological world have, and guide him/her in a aware use and interpretation of them.

Problems

Problem 3.1. Prove that the mean potential energy of vibration in a network of n nodes is

$$\langle V(\vec{x}) \rangle = \frac{1}{2n} \sum_{i=1}^{n} k_i R_i - \frac{1}{2n} \sum_{i,j \in E} (R_i + R_j - n\Omega_{ij})$$

where K_i and R_i are the degree and the sum of resistance distances of node i [72], and E is the set of edges of the network.

Problem 3.2. For the weighted network in Figure 3.13 assume that $\beta = 1$.

a) Write down the adjacency matrix
b) Calculate the vibrational centrality of each node.
c) Calculate the sum of resistance distances of each node.
d) Calculate the diversity of each node.
e) Calculate the centrality indices of points b), c) and d) in the case in which $w_{ij} = 1$ for all i, j.
f) Calculate the centrality indices of points b), c) and d) in the case in $\beta \neq 1$ (show some examples).
g) Consider an adjacency matrix defined in the following way [72]

$$H_{ij} = \begin{cases} \frac{1}{k_i^{out}}, & \text{if there is a link between i and j.} \\ 0, & \text{otherwise} \end{cases}$$

where k_i^{out} is the out-degree of node i. Could this network represent a Web page network? And biological signalling network? (*Hint*: make a reasoning on the role of node 4, whose out-degree is zero.)

Problem 3.3. For the transcription regulation network in Problem 2.4b) in Chapter 2, calculate the node entropy and the node vibrational centrality. Determine if there exists a relationship between these two indices.

Problem 3.4

a) Consider the transcription factor regulatory network of *E. coli* available at [54]. This is a fraction of the known network in *E. coli* where the regulated gene is also a transcription factor. Calculate the distribution of in- and out-degree, vibrational centrality, node entropy and node sum of resistance distance.

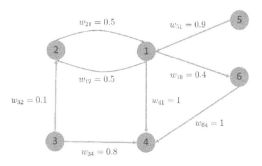

Figure 3.13: The network of Problem 3.2.

b) Consider the gene network of *E. coli* compiled by Shen-Orr et al. [196]. This is a gene transcriptional regulatory network formed both by transcription factors and target genes. Calculate the distribution of in- and out-degree, vibrational centrality, node entropy and node sum of resistance distance. Compare these distributions with those obtained for the transcription factor regulatory network at point a).

Chapter 4

Applied Descriptors for Complexity and Centrality to Network Biology

In this chapter, we discuss different network descriptors variously drawn in physics and how they can be utilized in order to help characterizing networks and their structural complexity. We start by surveying physical notions which have turned out of both theoretical and practical utility in network theory applications to systems biology and we outline their suitability to leverage network information for node prioritization. The focal part of the chapter describes entropy-based measures of structural complexity and outlines their suitability for the identification of distinct network components. The chapter proceeds describing network (dis-)similarity measures, firmly grounded in physics, that allow to analyse networks comparatively instead of individually. Finally, we conclude with the description of some of the most representative examples of applications of the introduced descriptors to biological networks for illustration.

4.1 Network Theory

Network science, and the diversity of theories developed therein, offers valuable approaches to construct and analyse living systems.

Current breakthroughs in high-throughput technologies spawns numerous efforts to decipher the data that are being acquired from complex systems on unprecedented scale, depth and complexity [168]. Transforming data production into meaningful biological insights remain very challenging. Owing to their generality, flexibility and power in pattern detection in virtually any relational structure, systems-oriented and graph theoretical approaches provide tools of unparalleled value for managing the complex nature, high dimensionality and compositional variability of biological systems [164]. Living systems have proven to be well represented and modelled as networks [7, 12, 21, 84]. Networks exist at all scales of biological organization, from genes that interact through mutual regulation to cells communicating with each other up to interacting species in ecosystems. A network conceptualization of a biological system consists of nodes, depicting system constituents (e.g. proteins in cellular networks or species in ecological networks), and edges depicting pairwise relationships between the system constituents (e.g. protein-protein binding interactions or feeding relationships).

A recurring challenge in complex systems analysis is the level of resolution of a network formalization. The amount of information retained in networks is determined by the study objective a priori. Network representations and their analyses range from being qualitative to highly quantitative. For example, living system analyses can include both uni-modal or multi-modal networks, where different types of nodes and/or edges could or could not be collapsed into a single type of nodes and edges, respectively. Binary undirected networks are the simplest representation where relationships between constituents are defined by the presence or absence of edges connecting the corresponding nodes. Approaches of intermediate complexity require more information than a binary representation but do not necessarily require quantifying system dynamics. Intermediate complexity approaches include qualitative models, which require only knowledge about the sign of interactions. In another intermediate approach, weighted networks incorporate the strength of interactions. Furthermore, probabilistic approaches, like Bayesian belief networks, express interactions between nodes as probabilities and contingencies. The most complex network analyses use dynamical system models where a set of ordinary differential equations describes interactions between nodes and require extensive parametrization.

Network theory studies in biology have reached a turning point, where empirical studies must provide the motivating details for novel theory and theoretical studies must provide a rigorous predictive framework in which to test hypothesis about network structure and the dynamics of the processes that take place on networks. A legacy of complex systems analysis is the realization that network connectivity and dynamics are inseparable. Through a large number of studies, it has been shown that informative measurements of network topological features allow networks to be qualitatively and quantitatively characterized and their structures to be related with the respective dynamics [172, 173]. The past two decades have witnessed substantial advances in this direction through a series of seminal findings that underlined the perils of ignoring network topology. One of those events traces back to the discovery that many real networks could not fulfil the simplest and most straightforward model proposed by Erdös and Rènyi, which assumes that the components of complex systems are randomly wired. Three prominent developments catalysed a burst of research on network local and global topological properties and their consequences in biological systems, from degree distributions to degree-degree correlations, motifs and the structure in communities [92, 224]: Watts and Strogatz's exploration of small-world networks [227], Barabàsi and Albert's discovery of scale-free networks whose node degree distribution follows a power law distribution [20], and the Girvan and Newmans identification of modular organization of real networks [92]. While random graphs, which are variants of the Erdös-Rènyi model, have been widely used as a benchmark for many modelling and empirical studies, it was not until the discovery of scale-free networks that the dynamical properties of networks became the focus of extensive network research [187, 217].

Considerable attention has been addressed at network robustness, a defining feature of living systems. Concisely stated, robustness refers to the capacity of the network to maintain specific functionalities in response to perturbations. For example, metabolic networks can continue to sustain life even after enzyme deficiencies [24] or can dynamically adapt to flux pattern modification [24], and gene regulatory networks continue to function upon loss-of-function mutations. The robustness of living systems has been attributed to functional redundancy [230] - multiple constituents with similar tasks - and to the distributed nature of networked systems (distributed robustness) whereby local failures rarely impact on functionality owing to systemic network compensation [222]. For instance, Albert et al. [9] demonstrated that robustness is shared by inhomogeneous networks where the connectivity distribution decaying as a power law provides a surprisingly high tolerance against random failures. Concurrently, Albert et al. [9] noticed that robustness against random failures comes at the expense of vulnerability to informed attacks. Indeed, removal of highly connected nodes decreases the ability of remaining nodes of communicating with each other. Owing to its broad applications, vulnerability of networks has received growing interest [30, 177] in different types of networks such as metabolic [98, 178] and neural system [211] networks. Due to their practical implications, it is worth remarking that new and interesting avenues in quantitative analysis of network vulnerability are being opened besides the earliest analysis of it in relation to node degree distribution [77, 102].

Notwithstanding the heterogeneity of contributions to the topic, it is undoubtful that the conceptual framework to study network responses to perturbations is firmly rooted in physics [8, 78, 203]. For instance, robustness has been recently characterized in terms of the entropy of the transition probabilities between nodes associated with the adjacency matrix [64]. Furthermore, some approaches to assess node vulnerability have been inspired to physical quantum and classical oscillator models [77].

4.2 Measures for Network Complexity and Centrality

Centrality measures have witnessed a growth of attention since they allow to characterize the structural complexity of any network of interest. Conceptually, they can be distinguished into local centrality measures [115], which determine the influence of individual nodes in signal distribution through a network, and global network centrality measures, which evaluate networks holistically with respect to their structural complexity. A wide list of computational methods has been proposed to capture network complexity and to prioritize network components. A reason for the plurality of measures for both purposes can be traced in the fact that neither the notion of network complexity nor that of centrality within networks are uniquely definable, since they depend on the problem under investigation. For instance, it is no surprise that node centrality measures can encompass indices based not only on node degree but also on the distance between nodes. Such is the case for betweenness centrality and closeness, to name a few that are well known, but also for the centrality which can be defined according to the resistance distance inspired by electrical network theory [125]. The objective of the chapter is not to present a comprehensive list of direct approaches to gauge network structural complexity. For this, specialized reviews of high quality and information content already exist. In this Chapter, we provide an overview of descriptors of global and local structural characteristics of networks, particularly focusing on approaches grounded in the physics domain and by categorizing them according to the underlying physics paradigm.

4.2.1 Network Entropy: A statistical mechanics and quantum physics definition

Among the variety of measures usefully introduced to describe the complexity present in networks, the formalizations of the concept of entropy has come a long way and have so far resulted into a plethora of tools to assess complexity. There is no a priori preferred notion of entropy in physical applications to network analysis and its specific choice appears to be purpose-dependent [43]. Historically, the entropy notion was introduced by Clausius, who postulated the second law of thermodynamics, and was clarified by Boltzmann. Suppose that a macroscopic system consists of a large number N of particles. We assume each of the particles is in one of the energy levels $E_1 < E_2 < \ldots < E_m$. The number of particles in the level E_i is N_i, so that $\sum_i N_i = N$ is the total number of particles. A macrostate of our system is given by the occupation numbers $N_1, N_2, \ldots, N_m$. The energy of a macrostate is $E = \sum_i^N N_i E_i$. A certain macrostate can be realized by many configurations of N particles. Those configurations are called microstates. We count the number of ways of arranging N particles in m energy levels such that each energy level shows $N_1, N_2, \ldots, N_m$ particles. There are:

$$\binom{N}{N_1, N_2, \ldots, N_m} := \frac{N!}{N_1! \ N_2! \ \ldots \ N_m!} \tag{4.1}$$

such ways. This multinomial coefficient is the number of microstates realizing the macrostate. Boltzmann called (4.1) the thermodynamic probability of the macrostate.

If we are interested into the thermodynamic limit N increasing to infinity, we use the relative numbers $p_i = \frac{N_i}{N}$ to label a microstate and we consider the average energy per particle $\frac{E}{N} = \sum_i p_i E_i$. To find the most probable macrostate, we wish to maximize (4.1) under a certain constraint. The Stirling approximation of the factorials gives

$$\frac{1}{N} \log \binom{N}{N_1, N_2, \ldots, N_m} = H(p_1, p_2, \ldots, p_m) + O\left(\frac{1}{N}\log N\right) \tag{4.2}$$

where

$$H(p_1, p_2, \ldots, p_m) := \sum_i -p_i \log p_i \tag{4.3}$$

If N is large then the approximation (4.2) yields that we can maximize (4.3) instead of (4.1). For example, maximizing (4.3) under the constraint $\sum_i p_i E_i = E/N$, we get

$$p_i = \frac{e^{-\lambda E_i}}{\sum_j e^{-\lambda E_j}} \qquad (4.4)$$

where the constant λ is the solution of the equation

$$\sum_i E_i \frac{e^{-\lambda E_i}}{\sum_j e^{-\lambda E_j}} = \frac{E}{N} \qquad (4.5)$$

Distribution (4.4) is known as the discrete Maxwell-Boltzmann law.

It was not until Shannon that equation (4.3) was firstly introduced as "uncertainty measure" or "information measure". More precisely, Shannon defined the information measure as the reduced entropy relative to the maximum entropy that can exist in a system characterized by the same number of elements:

$$I = H_{max} - H \qquad (4.6)$$

Most attempts to define the Shannon entropy on a network are based on the calculation of Shannon entropy of the probability distribution of some descriptors. For instance, Estrada and Hatano (Estrada, Hatano. Statistical mechanical approach to subgraph centrality in complex networks. Estrada et al. [75]) have defined the Shannon entropy of a network using a tight-binding Hamiltonian of the form $H = -A$. This entropy is based on the probability $p_j = \frac{e^{\lambda_j}}{Z}$ of finding the network in a state with energy equal to $-\lambda_j$ where λ_j is an eigenvalue of the adjacency matrix A and $Z = \sum_{j=1}^{N} e^{\lambda_j}$. Furthermore, a quantum-like entropy has been introduced. Indeed, the generalization of thermodynamic entropy to quantum mechanics by von Neumann is a milestone in the field. In quantum mechanics, probability distributions are encoded by density matrices. A density matrix ρ is a Hermitian and positive semi-definite matrix whose trace is equal to unity. The density matrix admits a spectral decomposition as

$$\rho = \sum_{i=1}^{N} \lambda_i |\psi_i\rangle\langle\psi_i|$$

for an orthonormal basis $\{|\psi_i\rangle\}$, where λ_i are non-negative eigenvalues which sum up to 1. The density matrix allows to define the von Neumann entropy by

$$S(\rho) = -Tr(\rho \, log_2 \, \rho) = -\sum_{i=1}^{N} \lambda_i log_2 \lambda_i,$$

i.e. it is equal to the Shannon entropy of the eigenvalues of the density matrix, where by convention $0 \ log_2 0: \ = \ 0$. More recently, inspired by calculation of entropy in quantum mechanics, (10.1103/PhysRevX.6.041062) introduced a connectivity-based matrix, which non-trivially depends on the combinatorial Laplacian matrix of the network, to define a spectral entropy that does not depend on the distribution of some network descriptors but on the network as a whole. Similar to the density matrix, such a matrix should be definite and symmetric; therefore $\rho = \frac{1}{Z}e^{-\beta H}$, with **H** a symmetric matrix with non-negative eigenvalues, Z and β real numbers, is a suitable candidate. Second, the eigenvalues of ρ must sum to unity, thus imposing the constraint $Z = \ Tr \ e^{-\beta H}$. A possible definition is $\rho = \frac{e^{-\beta L}}{Z}, Z = Tr \ e^{-\beta L}$ where **L** = **D** − **A** denotes the combinatorial network Laplacian, being **A** the adjacency matrix and **D** the diagonal matrix of node degrees. It is worth remarking that when $\beta = \ -1, \ Z = 1, \ \boldsymbol{H} = \boldsymbol{A}$ we recover the communicability matrix by Estrada. From the eigen-decomposition of the Laplacian matrix $\boldsymbol{L} = \boldsymbol{Q\Lambda Q}^{-1}$ where $\boldsymbol{\Lambda}$ is the diagonal matrix of Laplacian's eigenvalues, it can be proven that $Z = \ \sum_{i=1}^{N} e^{-\beta \lambda_i(L)}$ where $\lambda_i(\boldsymbol{L})$ indicates the ith eigenvalue of **L**. The spectral entropy of ρ can be written as

$$S(\rho) = - \sum_{i=1}^{N} \lambda_i(\rho)log_2\lambda_i(\rho)$$

It follows that

$$\lambda_i(\rho) = \frac{1}{Z}e^{-\beta\lambda_i(L)}$$

Provides the relationship between the eigenvalues of the density and the Laplacian matrices. Using the previous equation, the spectral entropy of the network G reduces to

$$S(G) = \ log_2 Z + \ \beta Tr[\boldsymbol{L\rho}].$$

4.2.1.1 *Partition-based network entropy descriptors*

Entropy was firstly applied to networks by Trucco [207] and Rashevsky [181]. These measures use an arbitrary network invariant X and an equivalence criterion α to induce partitions of X denoted by $X_1, X_2, \ldots,$ X_k. We infer distributions of X by considering the following scheme:

$$\begin{pmatrix} 1 & 2 & \ldots & k \\ |X_1| & |X_2| \ldots & |X_k| \\ p_1 & p_2 & \ldots & p_k \end{pmatrix} \tag{4.7}$$

The first row contains the equivalency classes and the second row the cardinalities of the partitions. The entities $p_i = \frac{|X_i|}{|X|}$ in the third row of the matrix can be used to infer probabilities for each obtained partition because it obviously holds $\sum_i p_i = 1$. Hence, $P_G = (p_1, p_2, \ldots, p_k)$ represents a finite probability distribution of G. Applying equation (4.3) leads to the graph entropy:

$$I(G, \alpha) = |X| \log(|X|) - \sum_{i=1}^{k} |X_i| \log(|X_i|) \tag{4.8}$$

Further studies explored the properties of this measure relative to different algebraic equivalence criteria like network automorphism and network colouring [158]. A limiting aspect of this measure relies on the fact that structurally non-equivalent networks may yield the same value. For instance, when the criterion of equivalence α is based on the equality of vertex degrees or orbits of the automorphism group of the network, it turns out that $I(G, \alpha) = 0$ both for complete and disconnected networks. However, it is undoubtful that a complete network is more complex than a disconnected one. This degeneracy issue is raised by the fact that the underlying symmetry is suitable to define compositional instead of topological complexity of a system.

This issue was overcome by a measure of network topological complexity introduced by Bonchev [35]. Assuming the network is represented by some kind of elements (e.g. vertex, edge, distance or clique) and that each element i is assigned a certain weight w_i, the probability of the element i of manifesting the weight w_i is defined by $p_i = \frac{w_i}{\sum_i w_i}$ with $\sum_i w_i = w$. Using Shannon's entropy, the probability scheme described by

$$\begin{pmatrix} 1, & 2, & N \\ w_1, & w_2 & w_N \\ p_1, & p_2 & p_N \end{pmatrix} \tag{4.9}$$

allows to define several measures. If the elements are vertices and the weights are the vertex degrees, this formalization allows to introduce the information content of the vertex degree distribution:

$$I_{vd} = \sum_{i=1}^{N} w_i \, log_2 \, w_i \tag{4.10}$$

I_{vd} clearly distinguishes between the null complexity of the disconnected network ($p_i = 0$) and the non-null ($I(w) = log_2 V$) complexity of the complete one ($p_i = \frac{1}{N}$). The Shannon's entropy of a graph is defined by the formula derived from equation (4.3):

$$H(w) = w \, log_{2 \, w} - \sum_{i=1}^{N} w_i \, log_2 w_i \qquad (4.11)$$

The maximum entropy is obtained when all $w_i = 1$.

4.2.1.2 Parametric network entropy descriptors

Determining vertex partitions associated with an equivalence relation can often be very costly. Alternative methods to derive non-partition-based complexity measures for networks have recently been proposed [60]. Instead of inducing network partitions to later obtain probabilities for such partitions, the network entropy definition here mainly relies on the assignment of a probability value to each vertex of a network by means of an informational functional. Therefore, the resulting measure is associated with the individual nodes over the network. Here, an informational functional f is a monotone positive mapping that captures some structural information on the network.

For a vertex i, we define the quantity

$$p_i := \frac{f_i}{\sum_{j=1}^{N} f_j} \qquad (4.12)$$

where f is a certain monotonic positive functional. The quantities p_i can be interpreted as vertex probabilities of the network since it holds $\sum_i p_i = 1$. We can therefore define the structural information content of the network as:

$$I_f(G) = - \sum_{i=1}^{N} \frac{f_i}{\sum_{j=1}^{N} f_j} \, log \left(\frac{f_i}{\sum_{j=1}^{N} f_j} \right) \qquad (4.13)$$

Examples of vertex informational functionals include:

$$f_i := \alpha^{c_1} |S_1(i, G)| + c_2 |S_2(i, G)| + \cdots + c_D |S_D(i, G)| \qquad (4.14)$$

or

$$f_i := c_1 |S_1(i, G)| + c_2 |S_2(i, G)| + \cdots + c_D |S_D(i, G)| \qquad (4.15)$$

$c_k > 0$, $1 \leq k \leq D$, $\alpha > 0$, D is the diameter of the network and S_j denotes the j-sphere of the vertex i:

$$S_j(i, G) := \{v \in V \mid d(v_i, v) = j\}, j \geq 1.$$

Starting from i, the cardinality of $S_j(i, G)$ denotes the number of vertices, which possess distance equal to j. The choice of α can be performed by using an information theoretic graph similarity measure, for instance the Kullback-Leibler divergence. The main idea for deriving these information functionals is based on the assumption that starting from an arbitrary vertex, information spreads out via shortest paths in the network. The parameters c_k can be used to weight structural characteristics or differences of the network in each sphere. It can be shown that the resulting measures generalize most of the aforementioned partition-based measures. So far, parameterised information functionals based on metrical properties of networks, that is, j-sphere cardinalities, have been proposed [61].

4.2.1.3   Classical random walk entropies

Network formalism allows to use entropy of the transition probabilities between nodes, with respect to dynamics such as random walks and self-avoiding random walks, in order to quantify the diversity of access of individual nodes and to rank their prominence in the global organization [155, 165, 199]. A random walk is usually described as a time-discrete process. The dynamics is formulated in terms of a transition matrix $P = \{\pi_{ij}\}$ being the entry π_{ij} the probability that a walker hops from node i to node j. For usual random walks, the transition matrix P is defined as $P_{ij} = \frac{a_{ij}}{d_i}$, where $d_i = \sum_{j=1}^{N} a_{ij}$ is the out-degree of the node i, i.e. the number of edges outgoing from node i. The dynamics of the random walk is monitored by the time evolution of a vector $v^t = \{v_i^t\}$ whose i-th component accounts for the probability that the walker is placed at node i at time t. Then, starting from some initial occupation probability, v^0, the dynamics of the walker follows the time-discrete rate equation:

$$v_i^t = \sum_{j=1}^{N} P_{ij}\, v_j^{t-1}$$

with $i = 1, \ldots, N$. By repeating the process indefinitely, the probability distribution of the walker converges to the stationary distribution which, by elementary Markov chain theory, can be shown not to depend on the initial state, if the graph is strongly connected and aperiodic. Indeed, the stationary distribution is defined by the dominant left eigenvector of the row stochastic, normalized adjacency matrix of the network $D^{-1}A$, where D is the diagonal matrix of outdegrees. A key example at the core of the most used search engine of the World Wide Web (Brin and Page. Comput Networks ISDN Syst. 30, 107. 1998), PageRank, relies on such random walks to rank nodes over the network. PageRank values of the nodes in the network are defined as the stationary distribution attained by the walker and reflect node centralities. The basic formulation of the PageRank algorithm is applicable only to strongly connected networks. Indeed, the existence and uniqueness of the stationary distribution is only guaranteed provided the transition matrix is irreducible and aperiodic, i.e. fulfils the Perron-Frobenius theorem. However, many real-world networks do not fulfil to this condition.

In order to satisfy this latter condition in any kind of network topology, the usual PageRank algorithm was modified to integrate a long-distance hopping probability. In this case, the probability for the random walker to move along edges is associated with a damping factor β. The value of β indicates the probability that the walker moves along edges, whereas the value of $1 - \beta$ indicates the probability that the walker jumps to a random node. Therefore, the usual dynamics of the walker is updated as follows:

$$v' = \beta Pv + (1 - \beta)e/n$$

where v' is the updated PageRank vector, N is the total number of nodes and e is a vector with all entries equal to 1. In the basic PageRank algorithm, all nodes have the same chance to be chosen as the destination by adding the term e/N. The personalized PageRank considers a vector e' to adjust node ranks as follows:

$$v' = \beta Pv + (1 - \beta)e',$$

For instance, e' can be set to the node weights so that a higher node weight means a higher rank of the corresponding node. Node ranking is performed according to the values stored in v', which quantifies the probability to find the walker on a specific node once the stationary probability distribution has been reached.

While PageRank is based on simple random walks, alternative centrality measures are based on the stationary distribution attained by Ruelle-Bowens random walks [63]

Very recently, the dynamical setting of network-based diffusion has been extended to the quantum domain [159]. In this framework, the potential of quantum random walks for ranking nodes over a network has started to be explored [112]. For instance, a recent approach relying on a master equation for the walker motion interpolating between purely quantum coherent dynamics and the classical diffusion has been shown to allow finding a stationary solution yielding a unique and reliable quantum rank algorithm [189].

4.2.1.4 Quantum walk-based centrality measures

Recent advances in quantum physics applications to communication and information processing have prompted a beneficial crossover with the science of complex networks. Quantum walks have established roles in quantum information processing [51]. In particular, they are central to quantum algorithms created to tackle database search [50, 197], as well as modelling of biological processes [126], just to name a few. Quantum walks are the quantum mechanical analogue of the well-known classical random walk. Any quantum process can be viewed as a single particle walk on a network. Single-particle quantum walks represent a universal model of quantum computation, meaning that any algorithm for a quantum computer can be translated into a quantum walk on a network. The computational advantage of algorithms based on quantum versus classical random walks has attracted wide interest into extending the spectrum of network descriptors (correlation, distance and centrality indicators), which are traditionally based on statistical mechanics, to the quantum domain [29], in both continuous and discrete time via coined walks.

Herein, we introduce some attempts to formalize quantum versions of the PageRank algorithm which are presently under active investigation [29, 79, 231]. Quantum versions of PageRank have been discussed in both discrete- and continuous-time quantum walk frameworks. For instance, the quantum formalization proposed by Paparo et al. [170] and extended by Loke et al. [136] uses the Szegedy discrete-time quantum walk to quantize the directed Markov chains encoded by the Google matrix. The quantum formalization introduced by Sànchez-Burillo et al. [189] uses a continuous-time quantum walk. It is worth mentioning the non-trivial problem faced by either discrete- or continuous-time quantum walk formulations to reconcile the directness of the network with the requirement for unitary and reversibility of the walk. The different solutions proposed will be highlighted describing individual quantum walk formulations of the PageRank algorithm.

Quantum PageRank algorithm by Szegedy discrete-time quantum walk

Quantization of the Markov chain underlying the classical PageRank algorithm is set up using the Szegedy's formalism of the discrete-time quantum walk in order to account for the connectivity structure and the directedness of the network [170]. The Szegedy's quantum walk takes place on the Hilbert space, which is the span of all vectors representing the $N \times N$ directed edges of the network, i.e. $H = \text{span}\{|i\rangle_1, |j\rangle_2\}$, with $i, j \in \mathbb{N} \times \mathbb{N}\} = \mathbb{C}^N \otimes \mathbb{C}^N$. The order of the spaces in the tensor product is crucial because we are dealing with a directed network. For each node j we introduce the quantum state vector

$$|\psi_j\rangle = \sum_{k=1}^{N} \sqrt{G_{kj}} \, |j\rangle_1, \, |k\rangle_2$$

This state is a weighted superposition of the quantum states representing the edges which are outgoing from the jth node whose weights are defined by the square root of the Google transition matrix G, i.e. $G = \beta P + (1 - \beta)E$, where P is the transition matrix and $E_{ij} = 1/(N-1)$ if $i \neq j$ while $E_{ii} = 0$. Unitary of the quantum walk is maintained since G is stochastic and information on the directionality is maintained in G. The quantum PageRank algorithm describes the quantum walk starting from the initial state $|\psi_0\rangle = \frac{1}{\sqrt{N}} \sum_{i=1}^{N} |\psi_j\rangle$. Its dynamics is determined by the quantum evolution operator $U = S(2\Pi - 1)$ where S is the swap operator, i.e. $S = \sum_{j,k=1}^{N} |j\rangle|k\rangle\langle k|\langle j|$ and Π is the projector operator, i.e. $\Pi = \sum_{j=1}^{N} |\psi_j\rangle\langle\psi_j|$. The ranking of the node N_i in the quantum network is based on the instantaneous quantum PageRank $I_q(N_i, t) = \langle\psi_0|U^{+2t}|i\rangle_2\langle i|U^{2t}|\psi_0\rangle$. In order to integrate out the fluctuations arising from the evolution, the average quantum PageRank is defined as the time-average of $I_q(N_i, t)$, i.e. $<I_q(N_i)> = \frac{1}{T} \sum_{t=0}^{T-1} I_q(N_i, t)$ which can be shown to converge for T sufficiently large. When the quantum PageRank algorithm was applied to diverse types of networks (scale-free, hierarchical and Erdös-Rényi random networks), it was found to improve the information gained on some properties of scale-free networks by allowing to: i) distinguish between different classes of networks, ii) to enhance the detection of hub nodes in scale-free networks, compared to the classical PageRank algorithm, by uncovering the presence of secondary hubs, and iii) to provide higher objectivity of the node rankings owing to its weaker dependence on the damping parameter β relatively to the classical algorithm. However, the quantum PageRank algorithm showed sensitivity toward coordinated attacks in scale-free networks, since the list of quantum PageRank values of the N nodes was found to change as a whole if the most important nodes fail due to hacker attacks and the algorithm is recalculated on the remaining nodes [171]. Furthermore, the quantum ranking algorithm demonstrated higher sensitivity under coordinated attacks compared to the classical one. This result is likely related to the augmented ability of the quantum algorithm to resolve structural details relative to the classical one. Said differently, when hubs are coordinately attacked, the order of less important nodes can change in the quantum case, where ranking values slightly change, whereas the order does not change in the classical case due to the degeneracy of a larger number of nodes.

Hybrid classical-quantum walk PageRank algorithm

The usual way of quantizing the Markovian random walk underlying the classical PageRank algorithm consists in introducing the Shrödinger equation

$$i\frac{h}{2\pi}\frac{d}{dt}|\psi(t)\rangle = H|\psi(t)\rangle$$

where the occupation probabilities are replaced by the components $\langle i\,|\,\psi\rangle$ of the wave function such that

$$\frac{d}{dt}\langle i\,|\,\psi\rangle = -\frac{i2\pi}{h}\sum_j \langle i\,|\,H|j\rangle\langle j\,|\,\psi\rangle$$

where $\langle i\,|\,H|j\rangle = P_{ij}$. It is worth noting that requiring unitary evolution operator in quantum mechanics implies that the H operator is Hermitian, i.e. $H = H^\dagger$, hence $\langle i\,|\,H|j\rangle = \langle j\,|\,H|i\rangle$. This procedure is thus restricted to undirected networks. Therefore, quantizing classical random walk is not straightforward since classical random walks are irreversible whereas quantum ones are reversible. To introduce irreversibility and quantum coherence, it was proposed [189] to write the Markovian master equation for the density matrix ρ rather than the Schrödinger equation. In this case, the mapping between quantum and classical random walks lies on the fact that the diagonal elements of ρ account for the occupation probabilities of each node. It turned out that the Markovian master equation can be casted in the form

$$\frac{d\rho}{dt} = -i(1-\alpha)[H,\rho] + \alpha\sum_{(i,j)}\gamma_{(i,j)}\left(L_{(i,j)}\rho L_{(i,j)}^\dagger - \frac{1}{2}\left\{L_{(i,j)}^\dagger L_{(i,j)},\,\rho\right\}\right)$$

where H is the Hamiltonian ($H = H^\dagger$) incorporating the quantum coherent dynamics, whereas the matrix γ, which describes the effect of non-energy-conserving phenomena such as temperature on the system, and the set of operators $\{L_{(i,j)}\}$, which forms a basis for the space ρ is embedded in, are responsible of irreversibility. The parameter $\alpha \in [0,1]$ quantifies the interplay between unitary and irreversible dynamics. For instance, the usual random walk can be incorporated by choosing $L_{(i,j)} = |i\rangle\langle j|$ and $\gamma_{(i,j)} = P_{ij}$. In this case, in the limit $\alpha = 1$, the equations for the diagonal elements, $\rho_{jj} = p_j$, meet those of the usual classical walk. Thus, this formulation interpolates between classical ($\alpha = 1$) and purely quantum random walks ($\alpha = 0$). We previously mentioned that the classical basic PageRank algorithm was modified by transforming the usual transition matrix P into the Google matrix G in order to guarantee a unique stationary distribution for any network topology. It turned out that the correction mechanism introduced in the classical case to ensure the existence of a unique stationary solution is generalizable to the quantum case by using the operators $L_{(i,j)} = |i\rangle\langle j|$ and $\gamma_{(i,j)} = G_{ij}$. Being linear, the time evolution for the master equation can be written in a compact form as $\frac{d\rho}{dt} = \mathcal{L}[\rho]$ which can be solved for any time t by taking the matrix exponential of $\mathcal{L}$, i.e. $\rho = \rho_0 e^{\mathcal{L}}$. The ranking of node N_i is given by $I(N_i) = \langle i|\rho|i\rangle$.

4.3   Comparative Network Analysis

The possibility of capturing and adequately quantifying (dis-)similarities between networks is a fundamental task in cutting-edge applications of network theory including but not limiting to model selection and comparative analysis of biological networks [17, 104, 111, 133]. In the past, it has been a common practice to assume that nodes are linked by a single type of static edge encapsulating all salient information. This assumption clearly oversimplifies the complexity of biological systems where nodes are linked by multiple relationships. More recently, multilayer approaches for network modelling, which account for different types of interactions between nodes, have provided a mathematically grounded framework to suitably represent the structural complexity featured by biological systems [124]. Multilayer networks consist of composite networks where nodes replicated across several networks show different types of connectivity. Each type of connectivity in multiplex networks constitutes the network layer. Multilayer networks can be used to model many biological systems [15, 57, 118]. At cellular level, basic functions are performed by the coordinated action of many layers of biomolecular interactions. At physiological level, it is well established that neuronal communication relies on multiple interaction modes including synaptic connections, gap junctions and neurohumoral signalling [27, 58]. Different layers of connectivity also arise naturally in ecology [174].

The ultimate aim of comparative network analysis is to capture and quantitatively gauge different system states which are represented by different networks. Many approaches to evaluate network similarity have been proposed, based on perfect and error-prone network matching. Exact network matching approaches seek to determine an isomorphism between any two networks whereas error-prone approaches can encompass network alignments [145], comparison of network grammars and metrics based on network transformations [40]. Furthermore, the quantum divergence (or equivalently quantum relative entropy) formalism allows to quantify (dis-)similarities between networks.

The quantum Rényi entropy can be used to define the quantum Rényi divergence

$$D_q(\rho\|\sigma) = \frac{1}{q-1} log_2 \, Tr(\rho^q \sigma^{1-q})$$

which reduces to the quantum Kullback-Leibler divergence

$$D_1(\rho\|\sigma) = Tr[\rho(log_2\rho - log_2\sigma)]$$

for $q \to 1$.

In general, such divergences are not symmetric and bounded, making difficult their application to network comparison.

An alternative measure is the quantum Jensen-Shannon divergence [144]

$$J_q(\rho\|\sigma) = S_q\left(\frac{\rho + \sigma}{2}\right) - \frac{1}{2}[S_q(\rho) + S_q(\sigma)]$$

where $\mu = \frac{\rho + \sigma}{2}$ is referred to as mixture matrix. It turned out that $J_q^{1/2}$ defines a true metric for $0 \le q \le 2$ and can be used as measure of distinguishability [56] or similarity [144].

4.3.1 Maximum Likelihood Estimation and Model Selection

For a certain model and its set of parameters, the likelihood function measures the probability of observing the data according to the model parameters. Therefore, it is sufficient to perform maximum likelihood estimation to obtain the parameters of the model that better reproduce the data, according to the model. Quantum divergence provides a framework to define a concept similar to the likelihood function for probability distributions in the context of density matrices. In the classical setting, assume that the empirical probability distribution P(x) is obtained from observing the N outcomes $\{x_i\}$ of a stochastic variable X. Let $Q(x; \theta)$ be a model to approximate P(x), which depends on the parameters set θ. In this context, the Kullback-Leibler divergence measures the information gain when the model $Q(x; \theta)$ is used to explain the observation P(x), and it can be written as

$$D(P\|Q) = \int dx \, P(x) \, log_2\frac{P(x)}{Q(x; \theta)} = -S(P) - \int dx \, P(x) \, log_2 Q(x; \theta)$$

If the model Q is plausible, there exists a value θ^* such that the divergence is minimum and we are interested in finding such a value by minimizing the divergence with respect to θ. We notice that the first term in the right-hand side does not depend on θ and therefore plays no role in the minimization procedure. By noting that

$$P(x) = \frac{1}{N} \sum_{i=1}^{N} \delta(x - x_i)$$

where δ is the Dirac function, the second term in the right-hand side of the previous equation reduces to

$$\int dx P(x) \log_2 Q(x; \theta) = \frac{1}{N} \sum_{i=1}^{N} \log_2 Q(x_i; \theta)$$

which is proportional to the negative log-likelihood function. Here, the pre-factor can be safely neglected during the minimization procedure. Therefore, minimizing the Kullback-Leibler divergence allows to maximize the log-likelihood function:

$$min_\theta \{ D(P \| Q) \} = max_\theta \{ \log_2 \mathcal{L}(x; \theta) \}$$

The proposed framework can be used to achieve a similar result in the case of density matrices. If ρ is the density matrix of an empirical network and if $\sigma(\theta)$ is a model for such a network, it turns out that, by starting from the quantum Kullback-Leibler divergence and by arguments similar to the classical case, the following equation holds:

$$min_\theta \{ D(\rho \| \sigma) \} = max_\theta \{ \mathrm{Tr}[\rho \log_2 \sigma(\theta)] \}.$$

By comparing the right-hand side of the previous two equations, we define the network log-likelihood function by

$$\log_2 \mathcal{L}(\theta) = \mathrm{Tr}[\rho \log_2 \sigma(\theta)] \tag{4.16}$$

where the likelihood function can be calculated by exploiting the properties of the matrix exponential as

$$\mathcal{L}(\theta) = e^{Tr[\rho \log_2 \sigma(\theta)]} = det(e^{\rho \log \sigma(\theta)}).$$

Besides obtaining a maximum-likelihood estimation of parameters, this result can be used to define an operative procedure for model selection, i.e. the identification of the model that best reflects the observed data out of a set of candidate models. One solution to this problem has been given by Akaike, who proposed the Akaike Information Criterion (AIC), by showing that the expected value of the relative cross-entropy term in the KL divergence equals the log-likelihood of the model given the data plus a penalizing constant term which accounts for the number of free parameters. The AIC is given by:

$$AIC = 2k - 2\log_2 \mathcal{L}(\theta^*) \tag{4.17}$$

where k is the number of parameters of the model and we plug Eq. (4.16) into Eq. (4.17) for application to quantum networks. Considering a set of models $M = \{M_1, M_2, \ldots, M_n\}$, with number of parameters $k_1, k_2, \ldots, k_n,$ respectively, the most suitable candidate to explain the data is the one being a trade-off between as small as possible divergence from the data and as small as possible number of parameters, i.e. the one whose AIC is minimum. Similarly, other model selection criteria can be extended from information theory to the complex network framework such as the Bayesian information criterion (BIC) defined by

$$BIC = k\log_2 N - 2\log_2 \mathcal{L}(\theta^*)$$

4.3.2 Reducibility of Multiplex Networks

As aforementioned, many biological systems can be represented by networks consisting of multiple layers. Although it has been proven that retaining such multi-dimensional information provides relevant insights into the underlying phenomena, it is practically useful to interrogate the multiplex network about the possibility of minimizing the number of layers while retaining all the salient aspects about the whole system. Indeed, the computation of even basic structural descriptors for multilayer networks, such as centrality and measures based on walks, scales superlinearly or even exponentially with the number of layers and can thus result unfeasible already for medium-sized networks.

Quantum divergence has proven useful to reduce structural redundancy in multilayer networks [56]. Let us assume that a multilayer network is represented by the set of $A = \{A^{[1]}, A^{[2]}, \ldots, A^{[M]}\}$, whose elements are the M × N adjacency matrices of the M layers of N nodes. It is possible to use the von Neumann entropy to quantitate the difference between the multilayer network (or a reduced configuration of it) and the network resulting from the aggregation of all layers into a monolayer. For this purpose, we define the von Neumann $H(A)$ of the multilayer network as the sum of the von Neumann entropies of its M layers, i.e.

$$H(A) = \sum_{j=1}^{M} h_{A^{[j]}}$$

where

$$h_{A^{[j]}} = - \sum_{i=1}^{N} \lambda_i^{[j]} log_2(\lambda_i^{[j]})$$

and $\lambda_i^{[j]}$ are the eigenvalues of the rescaled Laplacian matrix associated with the adjacency matrix $A^{[j]}$ of the jth layer. The von Neumann entropy thus explicitly depends on the number of M layers and on the structure of each layer. Consequently, it will change if we aggregate some of the layers of the original multilayer network into a reduced network. A special case is given by the one-layer network whose adjacency matrix is obtained by summing the adjacency matrices of all the M layers of A, i.e., $A = A^{[1]} + A^{[2]} + \cdots + A^{[M]}$ and whose von Neumann entropy is h_A. If we aggregate some of the M layers into a reduced multilayer network $C = \{C^{[1]}, C^{[2]}, \ldots, C^{[X]}\}$ with $X \leq M$ layers, where $C^{[i]}$ is either one of the adjacency matrices of the original layers or the sum of two or more of them. The entropy per layer of the multilayer network C can then be defined by:

$$\overleftarrow{H}(C) = \frac{H(C)}{X} = \sum_{j=1}^{X} h_{A^{[j]}}$$

Distinguishability between the multilayer network C and the aggregated network A can be defined through:

$$q(C) = 1 - \frac{\overleftarrow{H}(C)}{h_A}$$

The larger $q(C)$, the more distinguishable is the multilayer network C from the aggregated network A. The aim is to find $argmax\ [q(C)]$, i.e. the optimally reduced multiplex C_{max} yielding the maximum value of distinguishability from the aggregated network A. If we denote by M_{opt} the number of layers corresponding to the maximum value of relative entropy, we can define the reducibility of a multilayer network A as:

$$\chi(A) = \frac{M - M_{opt}}{M - 1}$$

which is the ratio between the number of reductions $(M - M_{opt})$ and the total possible number of potentially reducible layers (M-1). It is worth noting that $\chi(A) = 0$ if the network cannot be reduced, i.e. $M_{opt} = M$ whereas $\chi(A) = 1$ if $M_{opt} = 1$, i.e. the M layers are reduced into a single one.

To obtain the optimal configuration of layers, i.e. the configuration of layers maximizing $q()$, it would in principle be necessary to explore the space of all the possible partitions of the M layers which is computationally unfeasible. Alternatively, it has been proposed to adopt a greedy agglomerative hierarchical clustering algorithm which consists of the following stages: (i) computation of the quantum Jensen-Shannon divergence to estimate all pairwise layer dissimilarities, (ii) performing hierarchical clustering of the distance matrix, (iii) computation of the relative entropy of the dendrogram resulting from clustering, and (iv) selection of the partition maximizing $q()$.

4.4 Applications

The interaction of distinct components in biological systems naturally gives rise to complex networked structures. Networks have constantly been in the focus of research for the past decade, with considerable advances in the characterization of their structural properties. Here, we illustrate some examples of applications of the concepts introduced in the previous sections to systems biology for illustration.

4.4.1 Node Ranking in Biological Networks

Even if biological network constituents are primarily characterized on the basis of their individual biochemical action, the characterization of their contextual properties within networked structures is at least equally important. Remarkably, the application of topological centrality measures to various networks – regulatory and metabolic networks – has been a valuable guide for finding node [62, 117] or node sets such as the minimum dominating set protein [233] carrying out vital functional activities. One of the earliest attempts reported in the literature related centrality to indispensability in the yeast protein-protein interaction map [115]. Specifically, proteins involved in higher number of interactions were found to be more likely lethal upon their removal. More recently, a combined analysis of protein-protein interaction networks and functional profiles in *S. Cerevisiae* and *C. Elegans* demonstrated a positive association between the extent of individual contribution to network entropy and lethality [146]. Assessing node centrality is a remarkably challenging when analysing multilayer networks [59]. For instance, the pageRank algorithm has been successfully extended from monolayer to multilayer networks to identify core regulatory genes [163].

4.4.2 *Entropy-Based Estimation of Differentiation Potency*

The capability of quantifying the differentiation potency at the single-cell level is of paramount importance. Recently, it has been proposed that high cellular diversity underpins the undifferentiated state, with differentiated cells representing a more uniform state [85, 140]. At the single cell level, an undifferentiated cell is expected to show high network entropy since it has to afford the activation of many developmental choices. In contrast, a terminally differentiated cell would show a low entropy since a few pathways specific for the commitment fate are necessary. Since such a difference in entropy is expected to translate at the population level, network entropy has been tested for the capability to serve as a quantitative mark of the position of single cells [205] and cellular samples [18] in the global differentiation hierarchy. In these studies, the genome-scale gene expression profile of a sample (either a single cell or a bulk sample) is used to assign weights to the edges representing protein-protein interactions. If e_i and e_j are the normalized expression levels of nodes i and j, the edge weight w_{ij} is assumed to be approximated by $w_{ij} \sim e_i e_j$. If we envisage a random walker on this network where weights are normalized to ensure that the outgoing weights of node i sum to unity, the stochastic matrix is defined by:

$$p_{ij} = \frac{x_j}{\sum_{k \in Neig(i)} x_k}$$

where $Neig(i)$ denotes the neighbours of node i. The global network entropy rate is defined in terms of the stationary distribution π of the stochastic matrix P, trough: $S_R = \sum_{i=1}^{N} \pi_i S_i$ where S_i is the local network entropy for each node i out of N: $S_i = - \sum_{j \in Neig(i)} p_{ij} \log p_{ij}$. It can be shown that the maximal possible entropy rate is attained by a stochastic matrix defined by $p_{ij} = \frac{A_{ij} v_j}{\lambda v_i}$ where $A = \{A_{ij}\}$ is the adjacency unweighted matrix and v and λ are, respectively, its dominant eigenvector and eigenvalue. To ensure the entropy rate is bounded between 0 and 1, network entropy rate is scaled relative to this maximal value. Finally, by fitting a mixture of Gaussians to the normalized entropy rates of the cell population and using the Bayesian information criterion to estimate the optimal number of potency states and the state-membership probability of each cell, entropy was proven to discriminate between differentiated and undifferentiated cells [18, 205].

4.4.3 *Entropy-Based Hallmark of Cancer*

The notion of entropy has been applied to characterize disease phenotypes with the aim of discriminating between healthy and cancer cells. In contrast to the previous example where entropy served as a single quantitative correlate of cellular state, this study explores the discriminatory ability attained by local entropies. Specifically, [229] considered the local entropy defined at the single-node level and at the subnetwork level. Thus, the network entropy of a node i is defined by

$$S_i = -\frac{1}{\log k_i} \sum_{j \in \text{Neig}(i)} p_{ij} \log p_{ij}$$

where p_{ij} defines a stochastic matrix on the network and k_i is the node degree. The network entropy computed over paths of length larger than 1 is given by

$$S_N^{(2)} \propto -\sum_{ij} K_{ij}^{(2)} \log K_{ij}^{(2)}$$

where $K_{ij}^{(2)}$ fulfils to an approximate diffusion equation over the network allowing for paths of maximum length 2. The authors found that single-node entropy is increased in cancer cells compared to healthy ones and confirmed this observation in multiple cell types. Furthermore, this study showed that higher-order network entropies are still discriminatory, albeit weakly, and that increasing the order of network entropy is not informative. An explanation may be that the interesting changes associated with network entropy in cancer are localised to neighbours and nearest neighbours in the interaction network.

Problems

Problem 4.1. Consider an undirected graph $G = (V, E)$, with vertex set $V = \{1, \ldots, n\}$ and edge-set E. Let G be a connected graph, i.e. for every pair of vertices $i, j \in V$, there is a path between i and j in the graph. Let M be the adjacency matrix of G. Suppose we have a set E_1 of edges which we already know to be in the graph (so $E_1 \subseteq E$ and we do not have to query it). Let $G_{E_1} = (V; E_1)$ be the subgraph induced by only these edges, and suppose G_{E_1} is not connected, so it consists of $c > 1$ connected components. Call an edge $(i, j) \in E$ "connector" if it connects two of these components. Design a quantum algorithm that finds a "connector" edge with an expected number of $O(n/c)$ queries to M.

Problem 4.2. A classical discrete random walk on a line is a particular kind of stochastic process. The simplest classical random walk a line consists of a particle (often called "the walker") jumping to either left or right depending on the outcomes of a probability system (often called "the coin") with (at least) two mutually exclusive results, i.e. the particle moves according to a probability distribution.

Let $\{X_n\}$ be a stochastic process which consists of the path of a particle which moves along an axis with steps of one unit at time intervals also of one unit as in Figure 4.1. At any step, the particle has a probability p of going to the right and $q = 1p$ of going to the left. Assume that each step is modelled by a Bernoulli-distributed random variable [216] and the probability of finding the particle in position k after n steps and having as initial position $X_0 = 0$ is given by the binomial distribution $B_n = \frac{1}{2}(X_n + n)$. The expected number of steps before node j is visited, starting from node i is known as *hitting time*.

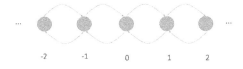

$$
\begin{array}{ccccc}
-2 & -1 & 0 & 1 & 2
\end{array}
$$

Figure 4.1: An unrestricted linear classical discrete random walk. The probability of going to the right is p and the probability of going to the left is $q = 1p$.

a) Write down the probability that the walker is in position k if he is in the position 0 at time 0:

$$\mathbb{P}(X_n = k|X_0 = 0).$$

b) Write down the distribution for the hitting time τ at some arbitrary point $k > 0$.

c) As described in Chapter 1, the exciton must propagate through chromophores in order to reach the reaction centre. Could this random walk model be suitably adapted to describe the exciton transfer in photosynthesis? *Hint:* it could be instructive to read Mohseni et al. [157].

Answers:

a)

$$P(X_n = k|X_0 = 0) = \begin{cases} \binom{n}{(n+k)/2} p^{\frac{1}{2}(n+k)} q^{\frac{1}{2}(n-k)} & \text{if } \frac{1}{2}(n+k) \in \mathbb{N} \cup \{0\} \\ 0 & \text{otherwise} \end{cases}$$

b)

$$P(\tau(k) = n) = \begin{cases} \frac{k}{n}\binom{n}{(n+k)/2}\frac{1}{2^n}, & \text{if } k+n \text{ is even} \\ 0 & \text{if } k+n \text{ is odd.} \end{cases}$$

Problem 4.3. At each step of a 1D-dimensional quantum walk, a particle can change its position to the superposition of right and left with the proper probability amplitudes gained by a quantum coin. Suppose that $\{|m\rangle, m \in \mathbb{Z}\}$ is the particle eigenstates constructing an orthonormal basis for the Hilbert space $\mathcal{H}_p$ of the particle [5]. The walk is driven by a coin, i.e. a 2D quantum system on a Hilbert space $\mathbb{H}_c$ generated by the orthonormal basis $\{|R\rangle, |L\rangle\}$. Each step of the quantum walk is given by the unitary

$$Q = (|R\rangle\langle R| \otimes S + |L\rangle\langle L| \otimes S^\dagger)(U_c \otimes \mathbb{I}))$$

acting on $\mathcal{H} = \mathcal{H}_c \otimes \mathcal{H}_p$. U_c defines the unitary operator of the coin and S and $S^\dagger$ are the shift operators:

$$S|m\rangle = |m+1\rangle$$
$$S^\dagger|m\rangle = |m-1\rangle$$

Starting from a pure coin-particle state $\psi(0)\rangle$, the state of the coin-particle at time t is [5]:

$$|\psi(t)\rangle = Q|\psi(0)\rangle.$$

a) Write down the expression of $|\psi(t)\rangle$ in the basis $\{|\psi_R(t)\rangle, |\psi_L(t)\rangle\}$.
b) Write down the particle density operator $\rho(t)$.

Answers:

a) $|\psi(t)\rangle = |R\rangle|\psi_R(t)\rangle + |L\rangle|\psi_L(t)\rangle.$

b) $\rho(t) = |R\rangle|\psi_R(t)\rangle\langle R|\psi_R(t)| + |L\rangle|\psi_L(t)\rangle\langle L|\psi_L(t)|$

Problem 4.4. Let $G = (V, E)$ be a graph with vertex set V (with $|V| = n$) and edge set E. The automorphism group of G, denoted by $\text{Aut}(G)$, is the set of all adjacency preserving bijections of V. Let $\{V_i | 1 \leq i \leq k\}$ be the collection of orbits of $\text{Aut}(G)$, and suppose $|V_i| = n_i$ for $1 \leq i \leq k$ [68].

a) Calculate the entropy I_A and I_B respectively of the graphs A) and B) in Figure 4.2. *Hint:* the orbits of graph A) in Figure 4.2 are $\{1\}, \{2, 5\}, \{3, 4\}$.

b) Give an example of graph with null entropy.

c) Which is the maximum value that the entropy of a graph with n vertices can assume?

Answers:

a) $I_A = \log 5 - \frac{4}{5} \log 2$, $I_B = \log 6$.

b) Every graph with the transitive automorphism group, such as the cycle C_n and the complete graph K_n on n vertices.

b) $\log n$.

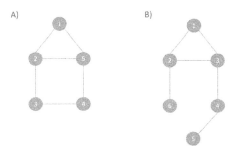

Figure 4.2: Examples on which to calculate the entropy of a graph.

Chapter 5

Perspectives

We conclude the book summarizing the motivations to support the convergence between theoretical physics and systems biology, tracing also some future perspectives and needs.

5.1 Systems Theory and Quantum Physics

Noteworthy progress has been made in applying concepts of systems theory [138] in biology resulting in the development of systems biology. Systems biology thrives on characterizing biological systems through integrated and quantitative model building on the basis of information gathered at molecular, cellular [53], multi-cellular and ecological community [174] scales. Biology addresses both the fine-grained characterization of mechanisms and the identification of generalizable themes helping to explain biological phenomena in different contexts. Current biology indeed embodies the realization that the two points of observations complement to each other in order to encompass the multifaceted answers necessary to relate information encoded in the singular components of a system to the behaviour of the system in its entirety in real-world conditions. Through half a century of experimentation in cell and molecular biology, ranging from labour-intensive and low-throughput studies focusing on single genes or gene products (e.g. biochemical characterization, perturbation) to high-throughput assays, biology has prominently assumed a reductionist mindset to explain biological systems according to the physical and chemical properties of their constituents [214]. Reductionism in particular can be distinguished into three types [81]: ontological reductionism, which assumes that everything that exists in nature is constituted by primitive elements that behave in a regular and predictable manner; epistemological reductionism, which argues that fundamental laws of a given level of organization can be derived from laws pertaining to a lower level; and methodological reductionism, which claims that complex phenomena can and should be understood by means of analysing their individual parts. In this view, the analysis of a biological system often occurs through the iteration of two basic operations: (i) system breakdown into components and (ii) extraction of connections among components by the integration of knowledge gathered on individual components.

Reductionism assumes that the phenomenology of a complex system can be explained in terms of simpler sub-system or by fundamental phenomena. Fast evolving multivariate biotechnologies such as genomics, transcriptomics, mass spectrometry proteomics, metabolomics and fluxomic data supported the methodological reductionism by accelerating the simultaneous measurements of hundreds to thousands of variables that make up many types of complex systems, over time and under various controlled or natural spontaneous perturbations. Technological advances have undoubtedly set the stage to transit towards systems molecular biology. As a matter of fact, biologists have been engaged into the organization, visualization, and synthesis of increasingly accumulating mounds of data to establish as much as possible comprehensive *in silico* models of manifold biological phenomena and systems. Many systems modelling studies do not necessarily end up in the derivation of definitive mechanisms but often provide a reduced experimental search space, or features prioritization or novel hypotheses, or they can enlighten discrepancies, errors or poorly characterized areas of knowledge that are worth of future investigation, in an iterative cycle of experiment, modelling and knowledge refining. In due course applying the reductionist approach, recently propelled by bewildering technical and analytical advances, has considerably improved our understanding of the mechanistic bases of living systems and has profoundly shaped our current mainstream intellectual framework.

The development of modern systems molecular biology has been paralleled by the augmented realization of the applicability of concepts drawn from systems theory to biological systems. Traditionally the study of complex biological systems has been dominated by statistics and dynamical models. In particular, the analysis of dynamical models (including phase plane analysis, stability, limit-cycle oscillation, bifurcation analysis, and sensitivity analysis) and has led to remarkable achievements such as the identification of recurrent themes across large numbers of models and model instantiations [139, 179], for instance in the topological organization of systems, their dynamics, responsiveness to structural and dynamical perturbations and controllability.

Although biology is intrinsically a science of complex systems, system-level abstraction of the complexity itself has only recently become prominent in the mathematical models. A system can be more than the sum of its parts, especially when it manifests synergies or *emergent properties*. Emergent properties are expressed by the system as a whole, whereas they are not expressed by its single parts. These properties arise because of the interactions among the parts of the system. Moreover, it is common knowledge that the properties of a biological system depend not only on the interactions among its components but also on external factors. The reductionist approach tends to remove the object of study from its context [195], in spite of the many cases in which this cause an oversimplification of the model for the system. For instance, the extrapolation of experimental results obtained under artificial and simplified conditions such as *in vitro* cell cultures could be misleading. Additionally, cancelling out the influence of environmental factors means neglecting adaptability which refers to the distinctive capacity of biological systems to react and modify environmental conditions in mutual interactions with each other. Another feature that is associated with reductionism and largely unmet in biological systems is the postulation of simple deterministic causations. The inapplicability of purely deterministic laws to biological systems firmly grounds on non-linear phenomena [2], functional redundancy [123] multiple components with similar functional effects - and pleiotropy a same component leading to multiple effects - which can be exerted by the individual components of a biological system.

The mounting appraisal of complexity of systems, particularly of living ones, requires us to recognize not only the opportunities but also the limitations of current means to study complex subjects. Scientists need to ground upon the realization of complexity in living systems to rethink the approaches to perform science and generate knowledge.

If biological phenomena kept to be a long-studied field by statistics and dynamical model theory, the theoretical physics domains (especially quantum and statistical mechanics) have largely been neglected to address the complexity of living systems, although theoretical physics concepts inspire deficient system-level models of phenomena. Furthermore, as a matter of fact, until now, the influence of quantum mechanics on biology has occurred in a reductionist sense to the extent that it describes the reality of inanimate matter on the tiniest scale, and, on the other hand, quantum effects were not expected to exert any role inside living systems. For years biologists and quantum physicists rarely crossed their paths. Biologists were wary of applying the world of theoretical (quantum) mechanics, where particles can be in more places at once or connected over vast distances or spread out like "ghostly" waves, to their own field, and physicists were reluctant to venture into the world of living systems, especially because up until a very recent past biology was not a quantitative science. Quantum physics and statistical mechanics can help to explain some natural phenomena we take for granted. As far as biology is probing the dynamics of ever-smaller systems, the signs of quantum mechanical behaviour in the building blocks of life are becoming increasingly apparent. Recent research suggests that some of life most fundamental processes can reach down to the quantum mechanics world to operate the extraordinary machinery of life. Though an overview of cases where quantum effects can be harnessed for assisting or enhancing biological functions is provided in the Section 1.7 of Chapter 1, here are a few examples. Advances in transition state theory together with the availability of structural, kinetic and thermodynamic data and with the hints offered by *in silico* simulations allowed to garner evidences for a contribution of quantum mechanical tunnelling to the generalized transmission coefficient in the rate constant for a reaction for biochemical reactions catalyzed by enzyme [86]. It has been conjectured that electron tunnelling might act as signalling mechanism in biological systems that rely upon ligand-receptor activation such as olfaction [169, 208]. Even if the possibility that a macroscopic sensing process could respond to fundamentally quantum effects is fascinating, determining a completed theoretical basis for the electron tunnelling contribution to molecular recognition is an ongoing process [101] paving the way to further investigations. Other possible quantum effects in living systems range from the magnetoreception of some avian species, that can capture variations occurring in the Earth magnetic field, to the observation of quantum coherent superpositions in photosynthetic energy capture.

Although it is recent, the convergence of quantum physics and biology gave birth to a new discipline: the quantum biology. It is important to note that the advances in quantum biology [126] open up the potential to technological breakthroughs. For instance, it has been shown that single amino acids can be identified exploiting electric currents originated from electron tunnelling currents induced by the diffusion of single amino-acid molecules through nanogap electrodes [166]. In another study, the control of light-matter interactions through the generation of photonic entanglement in green fluorescent proteins was exploited to develop quantum spectroscopic techniques in biomaterials [198].

Beyond the quantum phenomena that depend on the discreteness of the energy levels, arguably the most important quantum features for several applications to real-world networks have been entailed in quantum coherence and the role of environment on quantum coherence/decoherence. For instance, quantum communication has brought about a radical change in our understanding of the nature of information since quantum superposition and entanglement represent a type of parallelism. From the synergy witnessed between quantum mechanics and information processing, here we would like to stress the need for preserving phase relationships between different components of the wave function over a long enough duration for information to be processed, and the need for taking into account the properties and the influence of environment on the dynamical behaviour of a quantum system. Biological systems are definitely efficient information processors. Nonetheless, quantum mechanics is usually disregarded when we explore the information processing supported by biosystems. Here we argue that renewed attention to non-local quantum correlations, coherence, environment-assisted quantum transport [67, 183] or dephasing-assisted transport [52, 176], decoherence controllability [34] would precipitate novel opportunities for our understanding of biological information flow at biochemical and/or biological scales. Apart from their profound implications in our understanding of nature, the certification of non-local correlations could provide proof-of-principle for practical applications, most notably in the context of quantum biological and, possibly, non-biological phenomena. Furthermore, we believe that grounding the production of a conceptual representation of a biological system on the conceptual tools of quantum physics and the statistical mechanics describing it, such as the principle of state superposition and that of entanglement might result in substantial structural changes in network conception (e.g. concept itself of state variable) to allow addressing the known issue of identifiability from experimental data.

As a final remark, we emphasize the need for taking massive efforts in future research to innovate the existing approaches generating knowledge to accomplish the grand challenges ahead of us. In our opinion, one paramount objective is to prioritize the questions being asked rather than the tools being applied. Undoubtedly, a collaborative international, thoroughly balanced interdisciplinary effort in research is necessary. Furthermore, we envision that this innovation should be paralleled and energized by the development of the educational system in line with the expected changes in complex systems science.

5.2 Various Recapitulation Exercises

Problem 5.1. A gene X can be represented as a two state system, as it can be in a high expression state (ON) or in a low expression state (OFF). Let p the probability to be in the state ON and $1 - p$ the probability to be in the state OFF. Suppose that we make a large number N of independent observations of the state of the gene. We expect that in $m = Np$ of these N observations, the gene will be found in state ON, whereas in $N - m = N(1 - p)$, the gene will be found in state OFF.

a) Calculate the number $\mathcal{N}$ of different ways in which m observation can be in state ON, and $N - m$ observations can be in state OFF, among a total of N observations.

b) Write down the entropy H of the expression level of the gene in terms of $\mathcal{N}$.

c) Prove that from the expression of H found at point b) we can derive that

$$H = -p\log p - (1 - p)\log(1 - p)$$

where the choice of the base of the logarithm is matter of convenience, as it has the only effect to multiplying the entropy by a constant factor. *Hint:* use the Stirling's formula which states that $\log N! = N \log N - N$ for large values of N.

d) For which values of p, H is maximum? For which values of p, H is minimum? Give a plot of H vs p.

Answers:

a) $\mathcal{N} = \frac{N!}{m!(N-m)!}$.

b) $H = \frac{\log \mathcal{N}}{N}$.

d) H is maximized at $p = 1/2$, and is zero at $p = 0$ or $p = 1$.

Problem 5.2. Consider two genes X and Y. Since each gene is a two-state systems, the two-gene set is a four-state system (in general n genes form a 2^n-state system). Let $p_{i,j}$ be the probability that gene X is in state I and gene Y is simultaneously in state j. Let

$$H(X, Y) = -\sum_{i,j} p_{i,j} \log p_{ij}$$

be the joint entropy of X and Y.

a) Prove that if X and Y are independent, then

$$H(X,Y) = H(X) + H(Y).$$

b) Let $H(X|Y)$ and $H(Y|X)$ be the conditional entropies defined as

$$H(X|Y) = H(X,Y) - H(Y)$$
$$H(Y|Y) = H(X,Y) - H(X).$$

$H(X|Y)$ represents the uncertainty in X given the knowledge of Y (or, in other terms the residual uncertainty in X if Y were completely certain).

Prove that if X and Y are independent, then

$$H(X|Y) = H(X), \text{ and } H(Y|X) = H(Y).$$

c) Consider the mutual information between X and Y

$$M(X,Y) = H(X) - H(X,Y) = H(X) + H(Y) - H(X,Y),$$

which represent the amount of uncertainty in X if we remove the uncertainty associated with the knowledge of Y. Under which condition(s) $M(X,Y)$ is zero? Which is the probability that $M(X,Y)$ is zero? *Hint: $M(X,Y)$ can be expressed has the ratio of two maximum likelihoods, i.e. the maximum likelihood that the expression levels of X and Y are independent and the maximum likelihood that the expression levels of X and Y are arbitrary. Then, use the theorem that states that the distribution of -2 times the logarithm of the ratio of two likelihoods follows a χ^2 distribution.*

d) Prove that

$$M(X,Y) = \sum_{i,j}^{N} \log\left[\frac{p_{i,j}}{p_i p_j}\right],$$

[Answers: c) The expression levels of X and Y are independent.]

Problem 5.3. Imagine a gene regulatory network as in Figure 5.1 with three genes X, Y, and Z[1]. In such kind of networks, the probability distribution representing a node depends only on values taken by the random variables representing the parent nodes. $X(t), Y(t)$ and $Z(t)$

---

[1] This network is a *dynamic Bayesian network* [182]

represents the expression levels of the genes at time t. X regulates both Y and Z. The Figures shows that $X(t+1)$ is conditionally independent of $X(t-1)$ give $X(t)$ (i.e. the network satisfies the first-order Markov property). Write down the joint distribution of $X(t), Y(t), Z(t), X(t+1), Y(t+1), Z(t+1)$.

Answer:

$$P(X(t), Y(t), Z(t), X(t+1), Y(t+1), Z(t+1)) =$$
$$P(Y(t), Z(t)|X(t))P(Y(t+1), Z(t+1)|X(t+1))P(X(t+1)|X(t))P(X(t))$$

where $P(Y(t), Z(t)|X(t)) = P(Y(t)|X(t))P(Z(t)|X(t))$.

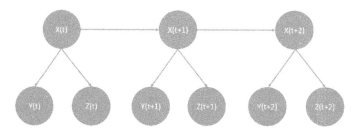

Figure 5.1: A gene regulatory network where the expression levels of genes Y and Z are conditionally dependent on the expression level of gene X at the same time point [182]. The distribution of Y at a particular time point, whereas the distribution of X at a particular time point is conditionally dependent on its value at the previous time point.

Problem 5.4. How to estimate a vector $\mathbf{x} \in \mathbb{R}^m$ from the measurement $\mathbf{y} = A\mathbf{x} \in \mathbb{R}^m$ where $A(m \times n)$ which $m < n$ is the measurement matrix (for example the value of n could be the number of genes, and m the number of samples of the measured expression level for each gene, in the case of gene expression network inference)? Assume, as in the majority practical cases, that $\mathbf{x} \geq 0$. In general, this task is often impossible as we have fewer measurements than variables. However, prove that if $\mathbf{x}$ is sparse, it can be calculated by solving the following problem[2],

$$\min \|x\|_0 \quad \text{such that} \quad Ax = y, \, \mathbf{x} \geq 0$$

where the L_0 norm $\| \cdot \|_0$ measures the number of non-zero entries of a given vector [223]. Prove that this method is NP-hard.

---

[2]The reader can deepen this subject by consulting the reference [223].

Problem 5.5. A *regular network* is a network in which every node has the same degree. Figure 5.2 simple example of regular networks. A 1-lattice is a special case of regular network in which each node is connected with its immediately neighbouring nodes in exactly the same way [182].

Consider a *random network* with some degree distribution p_k p_k is the probability that a node is chosen uniformly at the random from the network has degree k. Raval et al. [182] introduces the probability generating function

$$G_0(x) = \sum_{k=0}^{\infty} p_k x^k$$

such that

$$p_k = \frac{1}{k!} \frac{d^k G_0}{dx^k} \bigg|_{x=0}$$

and the average degree of a node is

$$\langle k \rangle = \sum_{k=0}^{\infty} k p_k = \frac{d G_0}{dx} \bigg|_{x=1}.$$

For an Erdös Rènyi random network of n nodes

$$p_k = \binom{n}{k} p^k (1-p)^{n-k}$$

where p is the probability that two nodes chosen at random are connected by an edge.

1-lattice

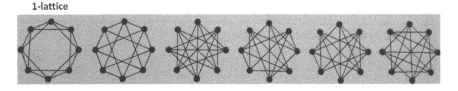

Figure 5.2: Regular networks of 8 nodes. In all these networks the node degree is 4. The first regular network is 1-lattice. Adapted from [182].

a) Implement a program to generate an Erdös Rènyi random network. Use it to generate six random networks with approximately the same number of edges of the regular networks in Figure 5.2. *Hint:* suitably set n and p in order to obtain approximately $\langle k \rangle$.

b) Calculate the entropy, the path length, and the clustering coefficient[3] of both the regular networks and the Erdös Rènyi random networks. Compare these centrality measures of the random networks with those of the random networks.

c) Which is the network with the lowest path length and the lowest clustering coefficient?

$$\left[\ \text{Answer: c) The 1-lattice.}\ \right]$$

Problem 5.6. Consider a graph and its equivalent electrical network like in Figure 5.3. The equivalent electric network of a graph is obtained by replacing an edge with a resistor. The resistance of this resistor is equal to reciprocal of the weight of the edge. The *effective resistance* is Ω_{ij} is defined as the voltage developed across a pair of nodes i and j when a unit current is injected at i and extracted from j, or vice versa. Let V_k^{ij} be the voltage of node k when a unit current is injected in i and a unit current is extracted from j. Prove that the expected number of times a random walk $i \to j$ visit node k can be expressed as $d_k V_k^{ij}$ where d_k is the degree of node k.

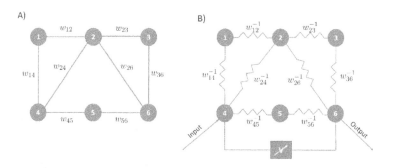

Figure 5.3: A) A graph and B) its equivalent electric network.

Problem 5.7.

a) Show that the topological centrality of a node in a network gives its overall position in the network (i.e. the distance of the node from the origin of the Euclidean space embedding the network.).

---

[3]The average clustering coefficient is a measure of the density of triangles (or cliques) in a network. In other terms it is the probability that the direct neighbours of a given node are connected to each other.

Hints: embed the network in a n-dimensional Euclidean space (n is the number of nodes to the network). This space is represented by the Moore-Penrose pseudo-inverse of the network Laplacian $\mathbf{L}^+$. The topological centrality of a node i is $C_i = \frac{1}{l_{ii}^+}$, where l_{ii}^+ are the diagonal elements of $\mathbf{L}^+$.

b) Show that Kirchhoff index

$$\mathcal{K} = n\,\mathrm{Tr}(\mathbf{L}^+)$$

determine the volume of the embedding. *Hint:* note that $\mathrm{Tr}(\mathbf{L}^+) = \sum_{i=1}^{n} \frac{1}{C_i}$.

Problem 5.8. Prove that graph entropy H is monotonic and sub-additive, according to the following definitions.

a) Monotonicity: f G_1 and G_2 are two graphs such that $V(G_1) = V(G_2)$ and $E(G_1) \subseteq E(G_2)$, where V denotes vertex set and E denotes edge set, then for any distribution probability P we have that

$$H(G_1, P) \leq H(G_2, P).$$

b) Sub-additivity: if G_1 and G_2 are two graphs on the same vertex set V, and $G_3 = G_1 \cup G_2$, then for any probability distribution P we have that
$$H(G_1 \cup G_2, P) \leq H(G_1, P) + H(G_2, P).$$

Using these results say if entropy give insights on a network structure. Could we use these results to say that entropy serves also as a measure of how structure of a network change with time? *Hints:* to answer to the second question, consider the various definitions of entropy, i.e. thermodynamic entropy, information entropy, and quantum entropy (Von Neumann entropy), and the links among them.

Problem 5.9. Only two of the following statements are true.

a) The quantum entropy of a network depends on the nodes' degree.

b) The entropy of a scale free network is smaller than the entropy of a random network.

c) The entropy of a network is a measure of the network's resilience to random failures.

d) Kirchhoff index does not reflect the robustness of a network.

e) A random network is a highly vulnerable.

$$\left[\text{Answer: a) and c) are true.} \right].$$

Problem 5.10. A protein can be considered as an elastic network of n C-α atoms, that are connected together by springs (chemical interactions).

a) Write down the potential energy V of the network under the harmonic oscillator approximation.

b) Express the potential energy V in terms of the Laplacian L of the network, and of the constant K of the elastic force. Which is the physical meaning of the eigenvalues of L?

Answers:

a) $V(x) = V(x = x_0) + \frac{1}{2}(x - x_0)^2 \frac{d^2 V}{dx^2} + \cdots$.

b) $V = \frac{K}{2}(\Delta x)^T L(\Delta x)$.

The vibrational normal modes of the protein are governed by the eigenvalues of L, and a small eigenvalue implies greater large-scale motion.

Problem 5.11. A difficulty of using mutual information M is that it takes values between 0 and infinity, turning the comparisons between different samples a difficult task. To overcome this problem it is common to use a standardized measure of mutual information, known in the literature as the *global correlation coefficient*, given by

$$\lambda = \sqrt{1 - \exp(-2M(X, Y))}.$$

a) Prove that $0 \leq \lambda \leq 1$.

b) Explain (also by examples) why λ captures also non-linear dependence between the variables X and Y.

Problem 5.12. The Hamiltonian of quantum harmonic oscillator is

$$H = \left(O^+ O + \frac{1}{2} \right) \hbar \omega, \quad O = \sqrt{\frac{m\omega}{2\hbar}} \left(x + \frac{ip}{m\omega} \right).$$

The coherent states ψ_c are defined as eigenstates of O. i.e.

$$O|\psi_c\rangle = \lambda|\psi_c\rangle.$$

a) Can we give the exact value of λ? *Hint:* consider the coherent state as a displaced harmonic oscillator ground state, and write $\lambda = \lambda_R + i\lambda_I$, $\lambda_{R,I}$ the real and imaginary parts of λ respectively. $\langle\psi|x|\psi\rangle = \sqrt{\frac{2\hbar}{m\omega}}\lambda_R$ and $\langle\psi|p|\psi\rangle = \sqrt{2m\hbar\omega}\lambda_I$, but the possible position in the $(x, 0)$ plane are infinite, so that there is no restriction on the values of $\lambda_{R,I}$.

b) Suppose to have a network of $2N+1$ coupled harmonic oscillators. Write down the Hamiltonian of the network. Explain the assumptions and the model you consider[4].

c) For which type of biological systems a (not necessarily harmonic) oscillators network model could be appropriate?

Answers:

a) No, λ is unbound.

b) Assuming that the mass of the oscillators is unitary, $H = \frac{1}{2}\left[\sum_{i=-N}^{N} p_i^2 + \sum_{i,k=-N}^{N} x_i A_{ik} x_k\right]$, where $\mathbf{A}$ is a $(2N+1) \times (2N+1)$ matrix characterizing the interactions among the oscillators. In giving this solution we do not make any particular assumption on $\mathbf{A}$ except that it is symmetric.

c) Two examples of networks whose nodes are oscillators could be neuronal networks, and networks governing biological clocks.

Problem 5.13. Consider a gene with two transcription factors A and B. The promoter has two non-overlapping operator site: O_A binds transcription factor A, and O_B binds transcription factor B.

a) In how many states can the promoter be found?'

b) Consider this reaction scheme

$$O + A \xrightarrow{k_1} O_A \qquad\qquad O_A \xrightarrow{k_2} O + A$$

$$O + B \xrightarrow{k_3} O_B \qquad\qquad O_B \xrightarrow{k_4} O + B$$

$$O_B + A \xrightarrow{k_1} O_{AB} \qquad\qquad O_{AB} \xrightarrow{k_2} O_B + A$$

$$O_A + B \xrightarrow{k_3} O_{AB} \qquad\qquad O_{AB} \xrightarrow{k_4} O_A + B$$

---

[4]The Hamiltonian expression depends on the assumption of the model of the set of oscillators, and in particular on the interactions among the oscillators.

with the following meanings: O: A and B unbound; O_A A bound at O_A, B unbound; O_B B bound at O_B, A unbound; O_{AB} A bound at O_A, B bound at O_B.

Simulate this reaction scheme with Gillespie algorithm for an arbitrary range of values for the rate constants k_1, k_2, k_3, and k_4. Analysing the results of the simulation identify which promoter state(s) would allow transcription if both A and B are repressors. *Hints:* from the simulations estimate the steady-state distribution of promoters (i.e. determine the fraction of promoters in all the possible states).

Answers:

a) In four states: O, O_A, O_B, and O_{AB}.
b) If both A and B are repressors and the binding of either factor prevents polymerase binding, then transcription can only occur from state O. If both A and B are repressors and the binding of both factors prevents polymerase binding, then transcription can only occur from state O_{AB}.

Problem 5.14. Consider a quantum network of N connected nodes, schematically shown in Figure 5.4. Each site is modelled as a qubit and may support an excitation which can be exchanged between lattice nodes by hopping. The initial (input) state is an excitation state which describes an excitation localized at node 1. The arrow between site N and sink denotes an irreversible transfer of excitations from site N to a sink. Let

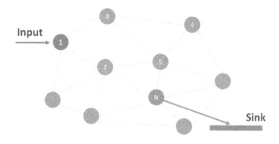

Figure 5.4: A quantum network of N connected nodes. This scheme of quantum network has been used to study the excitation energy transfer in photosynthetic complexes in literature [3, 121].

E_i be the energy of node i and w_{ij} a measure of the strength of the two-body coupling between ode i and node j.

a) Write down the Hamiltonian of the quantum evolution of this network using raising and lowering operators.

b) In a realistic scenario the nodes may suffer dissipative losses of energy as well as dephasing. Explain how/why.

Answers:

a) $H = \sum_{i=1}^{N} E_i O_i^+ O_i + \sum_{i \neq j} w_{ij}(O_i^+ O_j + O_j^+ O_i)$

b) Dissipative processes transfer the energy of node i to the environment. Dephasing process destroys the phase coherence of any superposition state in the system. The coherence caused by exitation decays over time, and the system returns to the state before exitation. Further reading about dephasing in photosynthesis is Engel et al. [70].

References

[1] G. Adesso, T.R. Bromley, and M. Cianciaruso. Measures and applications of quantum correlations. *Journal of Physics A: Mathematical and Theoretical*, 49(47):473001, 2016.

[2] M. Adler, P. Szekely, A. Mayo, and U. Alon. Optimal regulatory circuit topologies for fold-change detection. *Cell Systems*, 4(2):171–181.e8, 2017.

[3] B.-Q Ai and S.-L. Zhu. Complex quantum network model of energy transfer in photosynthetic complexes. *Physical Review E*, 86:061917, 2012.

[4] T. Aittokallio and B. Schwikowski. Graph-based methods for analysing networks in cell biology. *Mathematics Briefings in Bioinformatics*, 7(3Pp):243–255, 2006.

[5] S.J. Akhtarshenas, F. Shahbeigi, and A.T. Rezakhani. How quantum is a "quantum walk"? arXiv:1802.07027v1, 2018.

[6] J. Albert. A hybrid of the chemical master equation and the Gillespie algorithm for efficient stochastic simulations of subnetworks. *PLOS ONE*, 11(3):1–22, MAR 2016.

[7] R. Albert. Scale-free networks in cell biology. *Journal of Cell Science*, 118(21):4947–4957, 2005.

[8] R. Albert and A.L. Barabási. Statistical mechanics of complex networks. *Reviews of Modern Physics*, 74:47–97, Jan 2002.

[9] R. Albert, H. Jeong, and A.L. Barabási. Error and attack tolerance of complex networks. *Nature*, 406(6794):378–82, 2000.

[10] J. Almeida, M.B. Plenio, and S.F. Huelga. Origin of long-lived oscillations in 2d-spectra of a quantum vibronic model: electronic versus vibrational coherence. *The Journal of Chemical Physics*, 139:235102, 2013.

[11] G. Altay and F. Emmert-Streib. Revealing differences in gene network inference algorithms on the network level by ensemble methods. *Bioinformatics*, 26:1738–1744, 2010.

[12] L.A.N. Amaral and J.M. Ottino. Complex networks—augmenting the framework for the study of complex systems. *European Physical Journal B*, 38:147–162, MAR 2004.

[13] A. Ammar, E. Cueto, and F. Chinesta. Reduction of the chemical master equation for gene regulatory networks using proper generalized decompositions. *International Journal for Numerical Methods in Biomedical Engineering*, 28(9):960–73, 2012.

[14] K. Anand and G. Bianconi. Entropies of complex networks: toward an information theory of complex topologies. *Physical Review E (Rapid Communication)*, 80:045102, 2009.

[15] C. Angione, M. Conway, and P. Li. Multiplex methods provide effective integration of multi-omic data in genome-scale models. *BMC Bioinformatics*, 17 Suppl. 4:83, 2016.

[16] M. Arndt, T. Juffmann, and V. Vedral. Quantum physics meets biology. *Human Frontier Science Program Journal*, 3(6):386–400, 2009.

[17] S. Bandyopadhyay, M. Mehta, D. Kuo, M.K. Sung, R. Chuang, E.J. Jaehnig, B. Bodenmiller, K. Licon, W. Copeland, M. Shales, D. Fiedler, J. Dutkowski, A. Gunol, H. van Attikum, K.M. Shokat, R.D. Kolodner, W.K. Huh, R. Aebersold, M.C. Keogh, N.J. Krogan, and T. Ideker. Rewiring of genetic networks in response to DNA damage. *Science*, 330(6009):1385–9, 2010.

[18] C.R. Banerji, D. Miranda-Saavedra, S. Severini, M. Widschwendter, T. Enver, J.X. Zhou, and A.E. Teschendorff. Cellular network entropy as the energy potential in Waddington's differentiation landscape. *Scientific Reports*, 3:3039, 2013.

[19] M. Bansal, G. Della Gatta, and D. di Bernardo. Inference of gene regulatory networks and compound mode of action from time course gene expression profiles. *Bioinformatics*, 27(7):815–822, 2006.

[20] A.L. Barabási and R. Albert. Emergence of scaling in random networks. *Science*, 286(5439):509–12, 1999.

[21] A.L. Barabási and Z.N. Oltvai. Network biology: understanding the cell's functional organization. *Nature Reviews Genetics*, 5:101–113, 2004.

[22] R. Barbuti, A. Maggiolo-Schettini, P. Milazzo, A. Troina, and A. Troina. An alternative to Gillespie's algorithm for simulating chemical reactions. In *Computational Methods in Systems Biology (CMSB05)*, 2005.

[23] B. Barzel and A.L. Barabási. Network link prediction by global silencing of indirect correlations. *Nature Biotechnology*, 31(8):720–725, 2013.

[24] J. Behre, T. Wilhelm, A. von Kamp, E. Ruppin, and S. Schuster. Structural robustness of metabolic networks with respect to multiple knockouts. *Journal of Theoretical Biology*, 252(3):433–41, 2008.

[25] R.E. Bellman. *Dynamics programming*. Courier Corpo-ration, 2003.

[26] I. Bengtsson, S. Weis, and K. Zyczkowski. Geometry of the set of mixed quantum states: An apophatic approach. In P. Kielanowski, S. Ali, A. Odzijewicz, M. Schlichenmaier, and T. Voronov, editors, *Geometric Methods in Physics. Trends in Mathematics*, pages 175–197. Birkhäuser, Basel, 2013.

[27] B. Bentley, R. Branicky, C.L. Barnes, Y.L. Chew, E. Yemini, E.T. Bullmore, P.E. Vrtes, and W.R. Schafer. The multilayer *Connectome of Caenorhabditis* elegans. *PLoS Computational Biology*, 12(12):e1005283, 2016.

[28] M. Benzi. A note on walk entropies in graphs. *Linear Algebra and its Applications*, 455:395–399, 2013.

[29] S.D. Berry and Jingbo B. Wang. Quantum-walk-based search and centrality. *Phyical Review A*, 82:042333, Oct 2010.

[30] F. Bianconi, E. Baldelli, V. Ludovini, V. Luovini, E.F. Petricoin, L. Crin, and P. Valigi. Conditional robustness analysis for fragility discovery and target identification in biochemical networks and in cancer systems biology. *BMC Systems Biology*, 9:70, 2015.

[31] G. Bianconi. Entropy of network ensembles. *Physical Review E*, 79:036114, 2009.

[32] G. Bianconi, A.C.C. Coolen, and C.J. Perez-Vicente. Entropies of complex networks with hierarchically constrained topologies. *Physical Review E*, 78:016224, 2008.

[33] G. Bianconi, P. Pin, and M. Marsili. Assessing the relevance of node features for network structure. *PNAS*, 106:11433, 2009.

[34] D.N. Biggerstaff, R. Heilmann, A.A. Zecevik, M. Grfe, M.A. Broome, A. Fedrizzi, S. Nolte, A. Szameit, A.G. White, and I. Kassal. Enhancing coherent transport in a photonic network using controllable decoherence. *Nature Communications*, 7:11282, 2016.

[35] D. Bonchev and N. Trinajstić. Chemical information theory: structural aspects. *International Journal of Quantum Chemistry*, 22:463–480, 2009.

[36] C. Borgelt, M. Steinbrecher, and R. Kruse. *Graphical Models: Representations for Learning, Reasoning and Data Mining*, chapter Inductive Causation. Wiley Online Library, 2010.

[37] R. Borrelli and A. Peluso. Quantum dynamics of radiationless electronic transitions including normal modes displacements and duschinsky rotations: A second-order cumulant approach. *Journal of Chemical Theory and Computation*, 11(2):415–422, 2015.

[38] G.W. Brodland. How computational models can help unlock biological systems. *Seminars in Cell and Developmental Biology*, 47-48:62–73, 2015.

[39] J.C. Brookes. Quantum effects in biology: golden rule in enzymes, olfaction, photosynthesis and magnetodetection. *Proceedings. Mathematical, Physical, and Engineering Sciences*, 473(2201):20160822, 2017.

[40] H. Bunke and M. Neuhaus. Graph matching exact and error tolerant methods and the automatic learning of edit costs. In Diane J. Cook and Lawrence B. Holder, editors, *Mining Graph Data*. Wiley, 2006.

[41] J.M. Cai, F. Caruso, and M.B. Plenio. Quantum limit for avian magnetoreception: how sensitive can a chemical compass be? *Physical Review A Rapid Communications*, 85:040304(R), 2012.

[42] J.M. Cai, F. Caruso, and M.B. Plenio. Chemical compass for avian magnetoreception as a quantum coherent device. *Physical Review Letters*, 111:230503, 2013.

[43] M. Cai, Y. Cui, and H.E. Stanley. Analysis and evaluation of the entropy indices of a static network structure. *Scientific Reports*, 7(1):9340, 2017.

[44] R. Carlone, R. Figari, and C. Negulescu. The quantum beating and its numerical simulation. *Journal of Mathe-matical Analysis and Applications*, 450(2):1294–1316, 2017.

[45] F. Caruso, A.W. Chin, A. Datta, S.F. Huelga, and M.B. Plenio. Highly efficient energy excitation transfer in light-harvesting complexes: The fundamental role of noise-assisted transport. *The Journal of Chemical Physics*, 131, 2009.

[46] F. Caycedo-Soler, C.A. Schroeder, C. Autenrieth, A. Pick, R. Ghosh, S.F. Huelga, and M.B. Plenio. Quantum redirection of antenna absorption to photosynthetic reaction centers. *The Journal of Physical Chemistry Letters*, 8:6015–6021, 2017.

[47] V. Chaitankar, P. Ghosh, M.O. Elasri, K.A. Gust, and E.J. Perkins. Genome scale inference of transcriptional regulatory networks using mutual information on complex interactions. *BCB 2012 Proceedings of the ACM Conference on Bioinformatics, Computational Biology and Biomedicine*, pages 643–648, 2012.

[48] V. Chaitankar, P. Ghosh, M.O. Elasri, and E.J. Perkins. Time lagged information theoretic approaches to the reverse engineering of gene regulatory networks. *Bioinformatics*, 11(6):519, 2010.

[49] S. Chakraborty, L. Novo, S. Di Giorgio, and Y. Omar. Optimal quantum spatial search on random tempo-ral networks. *Physical Review Letters*, 119:220503, Nov 2017.

[50] A.M. Childs and J. Goldstone. Spatial search by quantum walk. *Phyical Review A*, 70:022314, Aug 2004.

[51] A.M. Childs, D. Gosset, and Z. Webb. Universal computation by multiparticle quantum walk. *Science*, 339(6121):791–4, 2013.

[52] A.W. Chin, A. Datta, F. Caruso, S.F. Huelga, and M.B. Plenio. Noise-assisted energy transfer in quantum networks and light-harvesting complexes. *New Journal of Physics*, 12:065002, 2010.

[53] L. Cowen, T. Ideker, B.J. Raphael, and R. Sharan. Network propagation: a universal amplifier of genetic associations. *Nature Reviews. Genetics*, 18(9):551–562, 2017.

[54] L. da F. Costa. Inward and outward node accessibility in complex networks as revealed by non-linear dynamics. https://arxiv.org/abs/0801.1982, 2008.

[55] K.C. Dasa, A.D. Güngörb, and A.S.C. Cevikb. On Kirchhoff index and resistance distance energy of a graph. *MATCH Communications in Mathematical and in Computer Chemistry*, 67:541–556, 2012.

[56] M. De Domenico, V. Nicosia, A. Arenas, and V. Latora. Structural reducibility of multilayer networks. *Nature communications*, 6:6864, 2015.

[57] M. De Domenico, S. Sasai, and A. Arenas. Integrating omic data with a multiplex network-based approach for the identification of cancer subtypes. *IEEE Trans Nanobioscience*, 15(4):335–342, 2016.

[58] M. De Domenico, S. Sasai, and A. Arenas. Mapping multiplex hubs in human functional brain networks. *Frontiers in Neuroscience*, 10:326, 2016.

[59] M. De Domenico, A. Sol-Ribalta, E. Omodei, S. Gmez, and A. Arenas. Ranking in interconnected multilayer networks reveals versatile nodes. *Nature Communications*, 6:6868, 2015.

[60] M. Dehmer. A novel method for measuring the structural information content of networks. *Cybernetics and Systems*, 39(8):825–842, 2008.

[61] M. Dehmer, K. Varmuza, S. Borgert, and F. Emmert-Streib. On entropy-based molecular descriptors: statistical analysis of real and synthetic chemical structures. *Journal of Chemical Information and Modeling*, 49(7):1655–63, 2009.

[62] G. del Rio, D. Koschtzki, and G. Coello. How to identify essential genes from molecular networks? *BMC Systems Biology*, 3:102, 2009.

[63] J.-C. Delvenne and A.-S. Libert. Centrality mea-sures and thermodynamic formalism for complex networks. *Physical Review. E, Statistical, Nonlinear, and Soft Matter Physics*, 83(4 Pt), 2011.

[64] L. Demetrius and T. Manke. Robustness and network evolution—an entropic principle. *Physica A: Statistical Mechanics and its Applications*, 346(3):682–696, 2005.

[65] K.N. Dinh and R.B. Sidje. Understanding the finite state projection and related methods for solving the chemical master equation. *Physical Biology*, 13(3):035003, 2016.

[66] K.N. Dinh and R.B. Sidje. An application of the Krylov-FSP-SSA method to parameter fitting with maximum likelihood. *Physical Biology*, 14(6):065001, 2017.

[67] R. Dutta and B. Bagchi. Environment-assisted quantum coherence in photosynthetic complex. *The Journal of Physical Chemistry Letters*, 8(22):5566–5572, 2017.

[68] F. Emmert-Streib. Entropy, orbits, and spectra of graphs. In *Analysis of Complex Networks: From Biology to Linguistics*, pages 175–197. WILEY-VCH Verlag GmbH & Co. KGaA, Weinheim, 2009.

[69] R.G. Endres. Making an impact in biology. *Physics Worlds*, pages 16–17, 2010.

[70] G.S. Engel. Quantum coherence in photosynthesis. *Procedia Chemistry*, 3(1):222–231, 2011. 22nd Solvay Conference on Chemistry.

[71] G.S. Engel, T.R. Calhoun, E.L. Read, T.-K. Ahn, T. Mancal, Y.-C. Cheng, R.E. Blankenship, and G.R. Fleming. Evidence for wavelike energy transfer through quantum coherence in photosynthetic systems. *Nature*, 446:782–786, 2007.

[72] E. Estrada. *The structure of complex networks. Theory and applications.* Oxford University Press, 2012.

[73] E. Estrada, J.A. de la Peña, and N. Hatano. Walk entropies in graphs. *Linear Algebra and its Applications*, 433:235–244, 2013.

[74] E. Estrada and H. Hatano. *Network Science. Complexity in Nature and Technology*, chapter Resistance distance, information centrality, node vulnerability and vibrations in complex networks, pages 13–29. E. Estrada, M. Fox, D. Higham and G.-L. Oppo, (eds), Springer, 2010.

[75] E. Estrada and N. Hatano. Statistical-mechanical approach to subgraph centrality in complex networks. *Chemical Physics Letters*, 439(1-3):247–251, 2007.

[76] E. Estrada and N. Hatano. Resistance distance, information centrality, node vulnerability and vibrations in complex networks.

In Ernesto Estrada, Maria Fox, Desmond J. Higham, and Gian-Luca Oppo, editors, *Network Science*. Springer, September 2010.

[77] E. Estrada and N. Hatano. A vibrational approach to node centrality and vulnerability in complex networks. *Physica A, Statistical Mechanics and its Applications*, 389:3648–3660, 2010.

[78] E. Estrada, N. Hatano, and M. Benzi. The physics of communicability in complex networks. *Physics Reports*, 514(3):89–119, 5 2012.

[79] M. Faccin, T. Johnson, J. Biamonte, S. Kais, and P. Migdał. Degree distribution in quantum walks on complex networks. *Phyical Review X*, 3:041007, Oct 2013.

[80] J.J. Faith, B. Hayete, J.T. Thaden, I. Mogno, and J. Wierzbowski. Large-scale mapping and validation of *Escherichia coli* transcriptional regulation from a compendium of expression profiles. *PLOS (Public Library of Science) Biology*, 5:54–66, 2007.

[81] F.C. Fang and A. Casadevall. Reductionistic and holistic science. *Infection and Immunity*, 79(4):1401–4, 2011.

[82] S. Feizi, D. Marbach, M. Mdard, and M. Kellis. Network deconvolution as a general method to distinguish direct dependencies in networks. *Nature Biotechnology*, 31(8):726–733, 2013.

[83] Z. Fox, G. Neuert, and B. Munsky. Finite state pro-jection based bounds to compare chemical master equation models using single-cell data. *The Journal of Chemical Physics*, 145(7):074101, 2016.

[84] L.I. Furlong. Human diseases through the lens of network biology. *Trends in Genetics : TIG*, 29(3):150–9, 2013.

[85] C. Furusawa and K. Kaneko. A dynamical-systems view of stem cell biology. *Science*, 2012.

[86] M. Garcia-Viloca, J. Gao, M. Karplus, and D.G. Truhlar. How enzymes work: analysis by modern rate theory and computer simulations. *Science*, 303(5655):186–95, 2004.

[87] D.T. Gillespie. A general method for numerically simulating the stochastic time evolution of coupled chemical species. *Journal of Computational Physics*, 22:403–434, 1976.

[88] D.T. Gillespie. Exact stochastic simulation of coupled chemical reactions. *Journal of Computational Physics*, 81(2340–2361), 1977.

[89] D.T. Gillespie. A rigorous derivation of the chemical master equation. *Physica A*, 188:404–425, 1992.

[90] D.T. Gillespie. Stochastic simulation of chemical kinetics. *Annual Review of Physical Chemistry*, 58(1):35–55, 2007.

[91] D.T. Gillespie. *Gillespie Algorithm for Biochemical Reaction Simulation*. Springer New York, 2013.

[92] M. Girvan and M.E. Newman. Community structure in social and biological networks. *Proceedings of the National Academy of Sciences of the United States of America*, 99(12):7821–6, 2002.

[93] K.-I. Goh, M.E. Cusick, D. Valle, B. Childs, M. Vidal, and A.L. Barabási. The human disease network. *PNAS*, 104(21):8685, 2007.

[94] M.C. Gonzales and A.L. Barabási. Complex network -from data to models. *Nature*, 3:224–225, 2007.

[95] A. Gupta, J. Mikelson, and M. Khammash. A finite state projection algorithm for the stationary solution of the chemical master equation. *The Journal of Chemical Physics*, 147(15):154101, 2017.

[96] H. Hache, C. Wierling, H. Lehrach, and R. Herwig. GeNGe: systematic generation of gene regulatory networks. *Bioinformatics*, 24(10):1318–1320, 2009.

[97] A.J. Hartemink. Banjo: Bayesian network inference with java objects. http://www.cs.duke.edu/~amink/software/banjo/.

[98] F. He, E. Murabito, and H.V. Westerhoff. Synthetic biology and regulatory networks: where metabolic systems biology meets control engineering. *The Federation of European Biochemical Societies (FEBS) Journal*, 13(117):20151046, 2016.

[99] M. Hegland, C. Burden, L. Santoso, S. MacNamara, and H. Booth. A solver for the stochastic master equation applied to gene regulatory networks. *Journal of Computational and Applied Mathematics*, 205:708–724, 2007.

[100] A. Hjartarson, J. Ruess, and J. Lygeros. Approximating the solution of the chemical master equation by combining finite state projection and stochastic simulation. In *Decision and Control (CDC), 2013 IEEE 52nd Annual Conference on*, 2013.

[101] R.D. Hoehn, D.E. Nichols, J.D. McCorvy, H. Neven, and S. Kais. Experimental evaluation of the generalized vibrational theory of g protein-coupled receptor activation. *Proceedings of the National Academy of Sciences of the United States of America*, 114(22):5595–5600, 2017.

[102] P. Holme, B.J. Kim, C.N. Yoon, and S.K. Han. Attack vulnerability of complex networks. *Physical Review. E, Statistical, Nonlinear, and Soft Matter Physics*, 65(5 Pt 2):056109, 2002.

[103] S. Hoyer, M. Sarovar, and K.B. Whaley. Limits of quantum speedup in photosynthetic light harvesting. *New Journal of Physics*, 12(6):065041, 2010.

[104] N.J. Hudson, A. Reverter, and B.P. Dalrymple. A differential wiring analysis of expression data correctly identifies the gene containing the causal mutation. *PLoS Computational Biology*, 5(5):e1000382, 2009.

[105] S.F. Huelga, A.W. Chin, and M.B. Plenio. Coherence and decoherence in biological system: Principles of noise assisted transport and the origin of long-lived coherences. *Philosophical Transactions of the Royal Society A*, 370, 2012.

[106] S.F. Huelga, M. del Rey, A.W. Chin, and M.B. Plenio. The role of non-equilibrium vibrational structures in electronic coherence and recoherence in pigment-protein complexes. *Journal of Physical Chemistry Letters*, 4:903–907, 2013.

[107] S.F. Huelga and M.B. Plenio. Vibrations, quanta and biology. *Contemporary Physics*, 54, 2013.

[108] S.F. Huelga, J. Prior, A.W. Chin, and M.B. Plenio. Efficient simulation of strong system-environment interactions. *Physical Review Letters*, 105, 2010.

[109] S.F. Huelga, J. Prior, A.W. Chin, and M.B. Plenio. The role of non-equilibrium vibrational structures in electronic coherence and recoherence in pigment-protein complexes. *Nature Physics*, 9:113–118, 2013.

[110] S.F. Huelga, C.A. Schröeder, F. Caycedo-Soler, and M.B. Plenio. Optical signatures of quantum delocalization over extended domains in photosynthetic membranes. *The Journal of Physical Chemistry A*, 119:9043–9050, 2015.

[111] T. Ideker and N.J. Krogan. Differential network biology. *Molecular Systems Biology*, 8:565, 2012.

[112] J.A. Izaac, Xiang Zhan, Zhihao Bian, Kunkun Wang, Jian Li, Jingbo B. Wang, and Peng Xue. Centrality measure based on continuous-time quantum walks and experimental realization. *Physical Review A*, 95:032318, Mar 2017.

[113] J. Jack, A. Păun, and A. Rodríguez-Patón. A review of the nondeterministic waiting time algorithm. *Natural Computing*, 10(1):139–149, March 2011.

[114] P.G. Jambrina, D. Herrez-Aguilar, F.J. Aoiz, M. Sneha, J. Jankunas, and R.N. Zare. Quantum inter-ference between h+d$_2$ quasiclassical reaction mechanisms. *Nature Chemistry*, 7:661–667, 2015.

[115] H. Jeong, S.P. Mason, A.L. Barabási, and Z.N. Oltvai. Lethality and centrality in protein networks. *Nature*, 411(6833):41–2, 2001.

[116] P.S. Jöberg. Numerical solution of the Fokker-Planck approximation of the chemical master equation. Master's thesis, Dept. of Information Technology, Uppsala University, 2005.

[117] A.C. Joseph and G. Chen. Composite centrality: A natural scale for complex evolving networks. *Physica D: Nonlinear Phenomena*, 267:58–67, 2014. Evolving Dynamical Networks.

[118] J.R. Karr, J.C. Sanghvi, D.N. Macklin, M.V. Gutschow, J.M. Jacobs, B. Bolival, N. Assad-Garcia, J.I. Glass, and M.W. Covert. A whole-cell computational model predicts phenotype from genotype. *Cell*, 150(2):389–401, 2012.

[119] P. Kaye, R. Laflamme, and M. Mosca. *An Introduction to Quantum Computing*. Oxford University Press, 2007.

[120] V. Kazeev, M. Khammash, M. Nip, and C. Schwab. Direct solution of the chemical master equation using quantized tensor trains. *PLoS Computational Biology*, 10(3):e1003359, 2014.

[121] N. Keren and Y. Paltiel. Photosynthetic energy transfer at the quantum/classical border. *Trends in Plant Science*, 23(6):497–506, 2018.

[122] A. Kimura and T. Kakitani. Theory of excitation energy transfer in the intermediate coupling case of clusters. *The Journal of Physical Chemistry B*, 107:14486–14499, 2003.

[123] H. Kitano. Towards a theory of biological robustness. *Molecular Systems Biology*, 3:137, 2007.

[124] M. Kivelä, A. Arenas, M. Barthelemy, J.P. Gleeson, Y. Moreno, and M.A. Porter. Multilayer networks. *Journal of Complex Networks*, 2(3):203–271, 2014.

[125] D.J. Klein and M. Randić. Resistance distance. *Journal of Mathematical Chemistry*, 12(1):81–95, 1993.

[126] N. Lambert, Y.-N. Chen, Y.-C. Cheng, C.-M. Li, G.-Y. Chen, and F. Nori. Quantum biology. *Nature Physics*, 9:10–18, 2013.

[127] P. Lecca. A stochastic description of the molecular mechanisms of hereditary parkinsonism. In *RECOMB '06*, 2005.

[128] P. Lecca. Simulating the cellular passive transport of glucose using a time-dependent extension of Gillespie algorithm for stochastic pi - calculus. *International Journal of Data Mining and Bioinformatics*, 1(4):315–335, 2007.

[129] P. Lecca, F. Bagagiolo, and M. Scarpa. Hybrid deterministic/stochastic simulation of complex biochemical systems. *Molecular BioSystems*, 13(12):2672–2686, 2017.

[130] P. Lecca, I. Laurenzi, and F. Jordan. *Deterministic versus Stochastic Modelling in Biochemistry and Systems Biology*. Woodhead Publishing Series in Biomedicine No. 21, 2012.

[131] P. Lecca, A. Re, A. Ihekwaba, I. Mura, and T-P. Nguyen. *Computational Systems Biology. Modelling and Inference*. Elsevier, 2016.

[132] K-C. Liang and X. Wang. Gene regulatory network reconstruction using conditional mutual information. *EURASIP Journal of Bioinformatics and Systems Biology*, 1:253894, 2008.

[133] Y. Lichtblau, K. Zimmermann, B. Haldemann, D. Lenze, M. Hummel, and U. Leser. Comparative assessment of differential network analysis methods. *Briefings in Bioinformatics*, 18(5):837–850, 2017.

[134] J. Lim, D. Palecek, F. Caycedo-Soler, C.N. Lincoln, J. Prior, H. von Berlepsch, S.F. Huelga, M.B. Plenio, D. Zigmantas, and J. Hauer. Vibronic origin of long-lived coherence in an artificial molecular light harvester. *Nature Communications*, 6:7755, 2015.

[135] S. Lloyd. Quantum coherence in biological systems. *Journal of Physics: Conference Series*, 302(1):012037, 2011.

[136] T. Loke, J.W. Tang, J. Rodriguez, M. Small, and J.B. Wang. Comparing classical and quantum pageranks. *Quantum Information Processing*, 16(1):25, Dec 2016.

[137] M. Lopes and G. Bontempi. Experimental assessment of static and dynamic algorithms for gene regulation inference from time series expression data. *Frontiers in Genetics*, 24(4):303, 2013.

[138] N. Luhmann. *Introduction to Systems Theory*. Polity Press, 2013.

[139] W. Ma, A. Trusina, H. El-Samad, W.A. Lim, and C. Tang. Defining network topologies that can achieve biochemical adaptation. *Cell*, 138(4):760–73, 2009.

[140] B.D. MacArthur and I.R. Lemischka. Statistical mechanics of pluripotency. *Cell*, 154(3):484–9, 2013.

[141] S. MacNamara, K. Burrage, and R.B. Sidje. Multiscale modeling of chemical kinetics via the master equation. *Multiscale Modeling and Simulation*, 6(4):1146–1168, 2008.

[142] S. MacNamara, R.B. Sidje, and K. Burrage. An improved dynamic finite state projection algorithm for the numerical solution of the chemical master equation with applications. *ANZIAM Journal*, 48(0):413–435, 2007.

[143] P.B. Madhamshettiwar, S.R. Maetschke, M.J. Davis, A. Reverter, and M.A. Ragan. Gene regulatory network inference: evaluation and application to ovarian cancer allows the prioritization of drug targets. *Genome Medicine*, 4:41, 2012.

[144] A.P. Majtey, P.W. Lamberti, and D.P. Prato. Jensen-shannon divergence as a measure of distinguishability between mixed quantum states. *Phyical Review A*, 72:052310, Nov 2005.

[145] N. Malod-Dognin, K. Ban, and N. Prulj. Unified alignment of protein-protein interaction networks. *Scientific Reports*, 7(1):953, 2017.

[146] T. Manke, L. Demetrius, and M. Vingron. Lethality and entropy of protein interaction networks. *Genome Informatics. International Conference on Genome Informatics*, 16(1):159–63, 2005.

[147] D. Manzano. Quantum transport in networks and photosynthetic complexes at the steady state. *PLOS ONE*, 8(2):1–8, 02 2013.

[148] D. Marbach, R.J. Prill, T. Schaffter, C. Mattiussi, D. Floreano, and G. Stolovitzky. Revealing strengths and weaknesses of methods for gene network inference. *Bioinformatics*, 107(14):6286–6291, 2010.

[149] A.A. Margolin, I. Nemenman I, K. Basso, C. Wiggins, G. Stolovitzky, R. Dalla Favera, and A. Califano. Aracne: an algorithm for the reconstruction of gene regulatory networks in a mammalian cellular context. *BMC Bioinformatics*, 20(7):1:57, 2006.

[150] D.A. McQuarrie. Stochastic approach to chemical kinetics. *Journal of Applied Probability*, 4:413–478, 1967.

[151] S. Menz, J.C. Latorre, Ch. Schtte, and W. Huisinga. Hybrid stochastic—deterministic solution of the chemical master equation. *SIAM Interdisciplinary Journal Multiscale Modeling and Simulation (MMS)*, 10:1232–1262, 2012.

[152] P.E. Meyer, K. Kontos, and F. Lafitte. Information-theoretic inference of large transcriptional regulatory networks. *EURASIP Journal on Bioinformatics and Systems Biology*, 79879, 2007.

[153] P.E. Meyer, F. Lafitte, and G. Bontempi. Minet: A r/bioconductor package for inferring large transcriptional networks using mutual information. *BMC Bioinformatics*, 9:461, 2008.

[154] L. Mirny. Leonid Mirny Course on Statistical Physics. https://ocw.mit.edu/courses/physics/8-592j-statistical-physics-in-biology-spring-2011/.

[155] L. Mirny. Leonid Mirny Laboratory. http://mirnylab.mit.edu/.

[156] M. Mohseni, Y. Omar, G.S. Engel, and M.B. Plenio. *Quantum Effects in Biology*. Cambridge Universiy Press, 2014.

[157] M. Mohseni, P. Rebentrost, S. Lloyd, and A. Aspuru-Guzik. Environment-assisted quantum walks in photo-synthetic energy transfer. *Journal of Chemical Physics*, 129:174106, 2008.

[158] A. Mowshowitz. Entropy and the complexity of graphs: I. an index of the relative complexity of a graph. *The Bulletin of Mathematical Biophysics*, 30:175–204, 1968.

[159] O. Muelken and A. Blumen. Continuous-time quan-tum walks: Models for coherent transport on complex networks. *Physics Reports*, 502:37–87, 2011.

[160] B. Munsky and M. Khammash. The finite state projection algorithm for the solution of the chemical master equation. *The Journal of Chemical Physics*, 124(4):044104, 2006.

[161] B. Munsky and M. Khammash. A multiple time interval finite state projection algorithm for the solution to the chemical master equation. *Journal of Computational Physics*, 226(1):818–835, 2007.

[162] B. Munsky and M. Khammash. A finite state projection algorithm for the stationary solution of the chemical master equation. *The Journal of Chemical Physics*, 147(15):154101, 2017.

[163] V. Narang, M.A. Ramli, A. Singhal, P. Kumar, G. de Libero, M. Poidinger, and C. Monterola. Automated identification of core regulatory genes in human gene regulatory networks. *PLoS Computational Biology*, 11(9):e1004504, 2015.

[164] M.E.J. Newman. The structure and function of complex networks. *Society for Industrial and Applied Mathematics (SIAM) Review*, 45(2):167–256, 2003.

[165] J.D. Noh and H. Rieger. Random walks on complex networks. *Physical Review Letters*, 92(11):118701, 2004.

[166] T. Ohshiro, M. Tsutsui, K. Yokota, M. Furuhashi, M. Taniguchi, and T. Kawai. Detection of post-translational modifications in single peptides using electron tunnelling currents. *Nature Nanotechnology*, 9(10):835–40, 2014.

[167] J. Omony. Biological network inference: a review of methods and assessment of tools and techniques. *Annual Research and Review in Biology*, 4(4):577–601, 2014.

[168] J.M. Ottino. Complex systems. *AIChE*, 49(2):292–299, 2003.

[169] M. Paoli, A. Anesi, R. Antolini, G. Guella, G. Vallortigara, and A. Haase. Differential odour coding of isotopomers in the honeybee brain. *Scientific Reports*, 6:21893, 2016.

[170] G.D. Paparo and M.A. Martin-Delgado. Google in a quantum network. *Scientific Reports*, 2:444, 2012.

[171] G.D. Paparo, M. Müller, F. Comellas, and M.A. Martin-Delgado. Quantum google in a complex network. *Scientific Reports*, 3:2773, 2013.

[172] R. Pastor-Satorras and A. Vespignani. Epidemic dynamics and endemic states in complex networks. *Physical Review. E, Statistical, Nonlinear, and Soft Matter Physics*, 63(6 Pt 2):066117, 2001.

[173] R. Pastor-Satorras and A. Vespignani. Epidemic spreading in scale-free networks. *Physical Review Letters*, 86(14):3200–3, 2001.

[174] S. Pilosof, M.A. Porter, M. Pascual, and S. Kfi. The multilayer nature of ecological networks. *Nature Ecology and Evolution*, 1(4):101, 2017.

[175] M. Pineda-Krch. Gillespiessa: Implementing the Gillespie stochastic simulation algorithm in R. *Journal of Statistical Software*, 25(12):1–18, 2008.

[176] M.B. Plenio and S.F. Huelga. Dephasing-assisted transport: quantum networks and biomolecules. *New Journal of Physics*, 10:113019, 2008.

[177] B. Podobnik, D. Horvatic, T. Lipic, M. Perc, Buldù, and H.E. Stanley. The cost of attack in competing networks. *Journal of the Royal Society Interface*, 12(112):20150770, 2015.

[178] M.J. Quinton-Tulloch, F.J. Bruggeman, J.L. Snoep, and H.V. Westerhoff. Trade-off of dynamic fragility but not of robustness in metabolic pathways *in silico*. *The FEBS Journal*, 280(1):160–73, 2013.

[179] S.J. Rahi, J. Larsch, K. Pecani, A.Y. Katsov, N. Mansouri, K. Tsaneva-Atanasova, E.D. Sontag, and F.R. Cross. Oscillatory stimuli differentiate adapting circuit topologies. *Nature Methods*, 14(10):1010–1016, 2017.

[180] G. Ranjan and Z.-L. Zhang. Geometry of complex networks and topological centrality. *Physica A: Statistical Mechanics and its Applications*, 392(17):3833–3845, 2013.

[181] N. Rashevsky. Life, information theory, and topology. *The Bulletin of Mathematical Biophysics*, 17:229–235, 1955.

[182] A. Raval and A. Ray. *Introduction to Biological Networks*. Taylor & Francis Group, CRC Press, 2013.

[183] P. Rebentrost, M. Mohseni, I. Kassal, S. Lloyd, and A. Aspuru-Guzik. Environment-assisted quantum transport. *New Journal of Physics*, 11:033003, 2008.

[184] T. Ritz, C. Damjanović, and K. Schulten. The quantum physics of photosynthesis. *Chemphyschem.*, 3(3):243–8, 2002.

[185] S. Roy, M. Werner-Washburne, and T. Lane. A system for generating transcription regulatory networks with combinatorial control of transcription. *Bioinformatics*, 24(10):1318–1320, 2009.

[186] O. Rozenblatt-Rosen, R.C. Deo, M. Padi, G. Adelmant, T. Rolland, M. Grace, A. Dricot, M. Askenazi, M. Tavares, S.J. Pevzner, F. Abderazzaq, D. Byrdsong, A.-R. Carvunis, A.A. Chen, J. Cheng, M. Correll, M. Durate, C. Fan, M.C. Feltkamp, S.B. Ficarro, R. Franchi, B.K. Garg, N. Gulbahce, T. Hao, A.M. Holthaus, R. James, A. Korkhin, L. Litovchick, J.C. Mar, T.R. Pak, S. Rabello, R. Rubio, Y. Shen, S. Singh, J.M. Spangle, M. Tasan, S. Wanamakter, J.T. Webber, J. Roecklein-Canfield, , E. Johannsen, A.-L. Barabási, R. Beroukhim, E. Kieff, , M.E. Cusick, D.E. Hill, , K. Munger, J.A. Marto, , J. Quackenbush, F.P. Roth, , J.A. DeCaprio, and M. Vidal. Interpreting cancer genomes using systematic host network perturbations by tumour virus proteins. *Nature*, 487:491–495, 2012.

[187] J. Runge, V. Petoukhov, J.F. Donges, J. Hlinka, N. Jajcay, M. Vejmelka, D. Hartman, N. Marwan, M. Palu, and J. Kurths. Identifying causal gate-ways and mediators in complex spatio-temporal systems. *Nature Communications*, 6(8502), 2015.

[188] V. Salari, H. Naeij, and A. Shafiee. Quantum interference and selectivity through biological ion channels. *Scientific Reports*, 7:41625, 2017.

[189] E. Sànchez-Burillo, J. Duch, J. Gmez-Gardees, and D. Zueco. Quantum navigation and ranking in complex networks. *Scientific Reports*, 2:605, 2012.

[190] K.R. Sanft and H.G. Othmer. Constant-complexity stochastic simulation algorithm with optimal binning. *The Journal of Chemical Physics*, 143(7):074108, 2015.

[191] A.S. Sanz and S. Miret-Artes. *Quantum Interference and Superposition*, chapter A Trajectory Description of Quantum Processes. II. Applications, pages 49–95. Springer-Verlag Berlin Heidelberg, 2014.

[192] T. Schaffter, D. Marbach, and D. Floreano1. GeneNetWeaver: *in silico* benchmark generation and performance profiling of network inference methods. *Bioinformatics*, 27(16):2263–2270, 2011.

[193] G.C. Schatz and M.A. Ratner. *Quantum Mechanics in Chemistry*. Dover Publications, Minnesota, NY, 2002.

[194] Serious Science. Statistical Phyiscs in biology. http://serious-science.org/statistical-physics-in-biology-1531.

[195] H. Shaked and C. Schechter. *Systems Thinking for School Leaders*, chapter Definitions and Development of Systems Thinking, pages 9–22. Springer, Cham, 2017.

[196] S.S. Shen-Orr, R. Milo, S. Mangan, and U. Alon. Network motifs in the transcriptional regulation network of *Escherichia coli*. *Nature Genetics*, 31:54–68, 2002.

[197] N. Shenvi, J. Kempe, and K.B Whaley. Quantum random-walk search algorithm. *Phyical Review A*, 67:052307, May 2003.

[198] S. Shi, P. Kumar, and K.F. Lee. Generation of photonic entanglement in green fluorescent proteins. *Nature Communications*, 8(1):1934, 2017.

[199] R. Sinatra, J. Gomez-Gardenes, Renaud Lambiotte, V. Nicosia, and V. Latora. Maximal-entropy random walks in complex networks with limited information. *Physical Review E*, 83:030103(R), 2011.

[200] R.J. Stanley, B. King, and S.G. Boxer. Excited state energy transfer pathways in photosynthetic reaction centers. 1. structural symmetry effects. *The Journal of Physical Chemistry*, 100(29):12052–12059, 1996.

[201] G. Stolovitsky and A. Califano. Dream 4: Dialogue for reverse engineering assessments and methods 4. http://www.the-dream-project.org/challenges.

[202] G. Stolovitsky and A. Califano. Dream: Dialogue for reverse engineering assessments and methods. http://www.the-dream-project.org/.

[203] Y. Sun, Y. Mao, and S. Luo. From quantum coherence to quantum correlations. *EPL (Europhysics Letters)*, 118(6):60007, 2017.

[204] V. Sunkara and M. Hegland. An optimal finite state projection method. *Procedia Computer Science*, 1(1):1579–1586, 2010. ICCS 2010.

[205] A.E. Teschendorff and T. Enver. Single-cell entropy for accurate estimation of differentiation potency from a cell's transcriptome. *Nature Communications*, 8:15599, 2017.

[206] M. Trautz. Das gesetz der reaktionsgeschwindigkeit und der gleichgewichte in gasen. best—"tigung der additivitt von cv-3/2r. neue bestimmung der integrationskonstanten und der molekldurchmesser. *Zeitschrift für Anorganische und Allgemeine Chemie*, 96(1):1–28, 1916.

[207] E. Trucco. On the information content of graphs: Compound symbols; different states for each point. *The Bulletin of Mathematical Biophysics*, 18:237–253, 1956.

[208] L. Turin, S. Gane, D. Georganakis, K. Maniati, and E.M. Skoulakis. Plausibility of the vibrational theory of olfaction. *Proceedings of the National Academy of Sciences of the United States of America*, 112(25):E3154, 2015.

[209] T. Udrescu and T. Jahnke. An adaptive method for solving chemical master equations using a sparse wavelet basis. In *AIP Conference Proceedings*, volume 489, page 1168, 04 2009.

[210] Northwestern University. Entangling biological systems. http://www.mccormick.northwestern.edu/news/articles/2017/12/entangling-biological-systems.html, 2017.

[211] M.P. van den Heuvel and O. Sporns. Network hubs in the human brain. *Trends in Cognitive Sciences*, 17(12):683–96, 2013.

[212] R. van Grondelle and V.I. Novoderezhkin. Quantum effects in photosynthesis. *Procedia Chemistry*, 3(1):198–210, 2011. 22nd Solvay Conference on Chemistry.

[213] N.G. van Kampfen. *Stochastic Processes in Physics and Chemistry.* Elsevier, Amsterdam, 1992.

[214] M.H. Van Regenmortel. Reductionism and complexity in molecular biology. Scientists now have the tools to unravel biological and overcome the limitations of reductionism. *EMBO Reports*, 5(11):1016–20, 2004.

[215] S. Vast, P. Dupont, Y. Deville, and P.S. Barbe. Automatic extraction of relevant nodes in biochemical networks. *7th Conference Francophone sur lapprentissage Auto-matique*, pages 21–31, 2005.

[216] S.E. Venegas-Andraca. Quantum information processing. *Quantum Information Processing*, 11(5):1015–1106, 2012.

[217] T. Verma, F. Russmann, N.A.M. Arajo, J. Nagler, and H.J. Herrmann. Emergence of coreperipheries in networks. *Nature Communications*, 7:10441, 2016.

[218] A.F. Villaverde, J. Ross, F. Morán, and J.R. Banga. MIDER: Network inference with mutual information distance and entropy reduction. *Plos One*, 9(5):e96732, 2014.

[219] H. Vo and R.B. Sidje. Improved Krylov-FSP method for solving the chemical master equation. In *World Congress on Engineering and Computer Science, Lecture Notes in Engineering and Computer Science*, volume 2226, pages 521–526, 2016.

[220] H. Vo and R.B. Sidje. An adaptive solution to the chemical master equation using tensors. In *The Journal of Chemical Physics*, volume 147, page 044102, 07 2017.

[221] H. Vo and R.B. Sidje. Solving the chemical master equation with the finite state projection and inexact uniformization in quantized tensor train format. In *5th International Conference on Computational and Mathematical Biomedical Engineering—CMBE, 2017*, page 1108, 04 2017.

[222] A. Wagner. Distributed robustness versus redundancy as causes of mutational robustness. *BioEssays : News and Reviews in Molecular, Cellular and Developmental Biology*, 27(2):176–88, 2005.

[223] M. Wang, W. Xu, and A. Tang. A unique nonnegative solution to an underdetermined system: From vectors to matrices. *IEEE Transactions on Signal Processing*, 59(3), 2011.

[224] X. Wang, E. Dalkic, M. Wu, and C. Chan. Gene module level analysis: identification to networks and dynamics. *Current Opinion in Biotechnology*, 19(5):482–91, 2008.

[225] X.-L. Wang, Q.-L. Yue, C.-H. Yu, F. Gao, and S.-J. Qin. Relating quantum coherence and correlations with entropy-based measures. *Scientific Reports*, page 12122, 2017.

[226] J. Watkinson, K.-C. Liang, X. Wang, T. Zheng, and D. Anastassiou. Inference of regulatory gene interactions from expression data using three-way mutual information. *Annals of the New York Academy of Sciences*, 1158:302–313, 2009.

[227] D.J. Watts and S.H. Strogatz. Collective dynamics of 'small-world' networks. *Nature*, 393(6684):440–2, 1998.

[228] L. Weber, W. Raymond, and B. Munsky. Identification of gene regulation models from single-cell data. *bioRxiv*, 2017.

[229] J. West, G. Bianconi, S. Severini, and A.E. Teschendorff. Differential network entropy reveals cancer system hallmarks. *Scientific Reports*, 2:802, 2012.

[230] J.M. Whitacre. Biological robustness: paradigms, mechanisms, and systems principles. *Frontiers in Genetics*, 3:67, 2012.

[231] J.D. Whitfield, C.A. Rodríguez-Rosario, and A. Aspuru-Guzik. Quantum stochastic walks: A generalization of classical random walks and quantum walks. *Phyical Review A*, 81:022323, Feb 2010.

[232] D.J. Wilkinson. *Stochastic Modelling for Systems Biology*. CRC Press, Taylor & Francis Group, 2012.

[233] S. Wuchty. Controllability in protein interaction networks. *Proceedings of the National Academy of Sciences of the United States of America*, 111(19):7156–60, 2014.

[234] Z. Xi, Y. Li, and H. Fan. Quantum coherence and correlations in quantum system. *Scientific Reports*, page 10922, 2015.

[235] M.A. Yildirim, K.-L. Goh, M.E. Cusick, A.-L. Barabási, and M. Vidal. Drug-target network. *Nature Biotechnology*, 25(10):1119–1126, 2007.

[236] X. Zhang, X.-M. Zhao, K. He, L. Lu, Y. Cao, J. Liu, J.-K. Hao, Z.-P. Liu, and L. Chen. Inferring gene regulatory networks from gene expression data by path consistency algorithm based on conditional mutual information. *Bioinformatics*, 28(1):98–104, 2012.

[237] P. Zoppoli, S. Morganella, and M. Ceccarelli. TimeDelay-ARACNE: Reverse engineering of gene networks from time-course data by an information theoretic approach. *BMC Bioinformatics*, 11(54), 2010.

Index

About the Authors

Paola Lecca has a Master Degree in Theoretical Physics and a PhD in Computer Science and Telecommunications. She worked for several years since its foundations as researcher and principal investigator at the Microsoft Research - University of Trento, Centre for Computational and Systems Biology, where she was the leader of the research team "Knowledge inference and data management". She is currently researcher at the Department of Mathematics of University of Trento, scholarship holder at the Department of Medicine of University of Verona, and Senior Professional Member of Association for Computing Machinery. Her main research activities focus on computational modelling and algorithmic procedures implementing efficient solutions for identiafiability, controllability, and simulation of complex dynamical networks. The main applicative domains of these studies are network biology, biochemistry, biological physics, microbiology, and synthetic biology. Dr. Paola Lecca is author of about hundred publications including books and journal and conference papers on international journals in computer science, computational biology, bioinformatics, and biophysics. She carries on an intense editorial activity as editor and reviewer for high impact journals in these subjects.

Angela Re obtained her Master degree in Physics in 2004 and Ph.D. degree in 2007 in the programme – Complex Systems Applied to Post-Genomic Biology – at the University of Turin. Since then, she has been conducting an active research program in international research centres, abroad and in Italy such as the Centre for Integrative Biology in Trento and the Centre for Sustainable Future Technologies in Torino. She has a long-proven track record working with mathematical conceptualization and statistical analysis of a variety of different biological data types. Specific areas of interest include the study of eukaryotic post-transcriptional regulatory mechanisms and their inclusion in cancer and stem cell pathways as well as the application of systems and synthetic biology approaches to bacterial RNA biology, proteomics and metabolism. She contributes to develop statistical software for the analysis of network modular organization and its dynamical properties, and she is interested into integrative multi-assay genomic data visualization and integration. She is in charge of editor ad reviewer activities in peer-reviewed scientific journals.

Milton Keynes UK
Ingram Content Group UK Ltd.
UKHW040052071024
449327UK00019B/500

9 780367 780388